The
Greatest Invention

Dung

Beetles

& A Cowman's Profits

THE
GREATEST INVENTION
DUNG
BEETLES
& A COWMAN'S PROFITS

CHARLES
WALTERS

Acres U.S.A.
Austin, Texas

Dung Beetles

Acres U.S.A.
P.O. Box 91299
Austin, Texas 78709 U.S.A.
(512) 892-4400 • fax (512) 892-4448
info@acresusa.com • www.acresusa.com

Printed in the United States of America

"The Ballad of the Scarab" by Euell Gibbons
used with permission of Rodale, Inc.

Publisher's Cataloging-in-Publication

Walters, Charles, 1926-
Dung beetles / Charles Walters. Austin, TX, ACRES U.S.A., 2008
    xxiv, 216 pp., 23 cm.
    Includes Index
    Includes Bibliography
    Includes Illustrations
    ISBN 978-1-60173-005-3 (trade)

    1. Dung beetles. 2. Dung beetles — ecology. 3. Dung beetles — nests. 4. Dung beetles — behavior. 5. Grazing — management. 6. Range ecology. 7. Pasture ecology. 8. Organic farming. 9. Agricultural ecology. I. Walters, Charles, 1926- II. Title.

    QL596.S3 W11 2008   595.76

*Dedicated to three young
ladies with inquiring minds —
Emily, Diana and Kara, my grandchildren.
It will be the younger generations
who best appreciate nature's resources from
the ground up, including nature's most
enduring animals, the beetles.*

Contents

"The Ballad of the Scarab" .... *ix*

Foreword .... *xiii*

Prologue.. *xvii*

1. The Greatest Invention 1

2. A Gift from Africa 13

3. From Down Under 29

4. A Lone Hetman... 37

5. The Call of the Beetle 45

6. Get to Know Truman Fincher 57

7. The Research Drill 69

8. The Dwellers.. 79

9. The Rollers... 83

10. The Tunnelers 89

11. The Klepto-Parasites.. 95

12. y-Sex... 99

13. Cast of Characters... 113

14. The Global Presence 127

15. Soils & Pastures 135

16. Five-Star Dining.. 141

17. Scanning the Literature 151

18. Competition.. 161

19. The Greatest Invention Revisited.................................. 167

20. A Codicil of Sorts.. 179

Afterword 187

Appendix 1: A Dung Beetle Demonstration Farm 197

Appendix 2: A Poem by Mark Sturges.................................. 201

Acknowledgment 205

Index 207

The Ballad of the Scarab

Let's take a trip to Egypt's land
And sing about the beetle,
Who rolls about upon the sand
A ball of something fecal.

Despite this lowly job of his,
He's not without some fame,
And *Scarabaeous sacer* is
His awesome Latin name.

Egyptians did appreciate
This tumblebug so nimble,
And by a vote of twelve to eight
Made him the sun-god's symbol.

Ten thousand images of him
In copper, brass, and gold,
And some in stone and precious gem
Were made by priests of old.

The scarab found a cozy nook
Within a temple attic,
There saw his picture in a book;
It made him feel ecstatic.

No wonder he was feeling smug,
The honor was terrific;
They'd made this humble tumblebug
Into a hieroglyphic.

"A mere dung beetle," you surmise
About the scarab small;
You cannot judge him by his size,
That will not do at all.

You can't dismiss him with a shrug;
Don't ever try to flout him,
For though he is a tumblebug,
There's nothing mere about him.

The scarab will ignore your sneers
He's proof against your scorn;
He'd been around a million years
When history was born.

Before the pyramids were raised
Upon the desert sand,
Egyptians sought the scarab's aid
To sanitize the land.

When Moses led his people out
To found the Hebrew nation,
The little scarab was about
To help with sanitation.

He crept into the private room
Of Cleopatra fair;
He saw historic romance bloom,
Mark Antony was there.

The scarab heard him threaten her
To conquer with his might,
Heard Cleopatra answer, "Sir,
I am not prone to fight."

He saw the Prophet's horses come,
Each mounted by an Arab;
It was a time most wearisome
For patient little scarab.

He had to work both night and day,
'Twas painful to endure;
It took a week to clear away
The piles of horse manure.

He later watched a Moslem village
The day Crusaders came;
He saw them murder, rape, and pillage,
All in religion's name.

The conquerors have come and gone,
And each has left his blot,
But still the scarab's rolling on
The balls of you-know-what.

And so from Cairo to Sudan
He worked from year to year
As ancient Egypt's only san-
Itary engineer.

He had to labor all the while,
Or very soon, methinks,
The smell you'd whiff along the Nile
Would surely rhyme with Sphinx.

When cause of illness was unknown,
And knowledge very vague,
The gallant scarab worked alone
To save men from the plague.

Had one ancestor caught typhoid
And childless passed away,
Then you'd be nothing but a void;
You'd not be here today.

So whether you are Greek or Jew,
Or whether Dutch or Arab,
The chances are you'd not be you
If there had been no scarab.

Let's give the scarab one big cheer
And never do him wrong;
It's by his work that we are here
To sing this silly song.

— *Euell Gibbons,*
in Organic Gardening and Farming, *May 1968*

Foreword

When asked what he could infer about the attributes of God, the British geneticist J.B.S. Haldane thought a moment and then remarked, "God must have an inordinate fondness for beetles since he created so many different species." At last count about 350,000 species of beetles have been described, and as many as 4-8 million species of beetles probably still remain to be described. Of all these beetles, it is the dung beetles that are described with inordinate fondness on the pages of this book. To the ancient Egyptians the dung beetle was the revered symbol of regeneration and renewal. A fondness for dung beetles permeates the 19th century writings of Jean Henri Fabre, that passionate observer of insect life. Fabre spent over forty years studying dung beetles, and his published works provided the first descriptions of the domestic lives of these buriers of dung.

Today, the presence of dung beetles on farms and ranches should be welcomed with the reverence of the ancient Egyptians and with the enthusiasm of Fabre. Dung beetles not only directly benefit agricultural ecosystems in myriad ways but also their beneficence extends indirectly to global climate and the health of our populations. Dung beetles have been estimated to save farmers $2 billion each year, and the savings in health costs and environmental degradation that can be attributed to these beetles are even far greater.

In these pages the reader will learn about dung beetles and the humans of the 20th century who have been their fondest champions. George Bornemissza of Australia and Truman Fincher of the United States are two scientists who have been among the most enthusiastic proponents of dung beetles and their great economic value to agriculture, the environment, and health. These individuals have carried out the carefully designed and controlled experiments that have unequivocally demonstrated the many unsung virtues of dung beetles. Other scientists continue to confirm and extend their findings while many farmers and ranchers have implemented these findings on their pastures and ranches.

Few people realize that dung is actually a resource that is coveted and fought over by many creatures. Manure happens to be the breeding ground of not only dung beetles, but also many pests and parasites of livestock. The insect pests include those infamous face flies, horn flies, and buffalo flies. Ninety-five percent fewer horn flies emerged from cowpats consumed by dung beetles than from cowpats from which dung beetles had been excluded. Parasites of the digestive tract of livestock are sometimes present in dung, but their rapid burial by dung beetles eliminates these parasites from a pasture and saves substantial veterinary costs. A dramatic nine-fold decrease in infections of parasites of livestock has been shown to accompany the arrival of dung beetles to pastures.

As a reflection of their discriminating tastes, most livestock avoid eating any forage in the immediate vicinity of their dung pats. Without efficient removal of dung from a pasture, areas covered by dung remain untouched and inefficiently used by livestock. Cattle will not graze on the rank growth of grass surrounding undigested dung. Removal of these pats by beetles can significantly improve the efficiency of livestock grazing.

Continual surfacing of inconvenient truths about causes and cures of global warming further enhances the status of dung beetles as agents that counter global warming. According to a 2007 report of the Food and Agriculture Organization of the United Nations, livestock are significant contributors to global warming, being responsible for 18 percent of greenhouse gas emissions. Livestock are said to be responsible for 9 percent of all CO_2 emissions, 37 percent of all methane (CH_4) emissions, and 65 percent of all nitrous oxide (N_2O) emissions. These sobering percentages actually exceed the percentage of emissions derived from sources of transportation. But the participation of dung beetles in rapidly incorporating nitrogen and carbon nutrients of dung into the soil can assist us in decreasing the release of these greenhouse gases from dung and dramatically reducing global warming.

According to the U.S. Environmental Protection Agency, livestock raised on nutrient-rich pastures with high quality forage produce less methane than grain-fed livestock raised in the "bovine concentration camps" represented by many livestock feed lots. The droppings of livestock in these concentrated animal feeding operations are soon pounded into hard slabs that cannot be utilized by even the most robust dung beetles.

However, in a pasture with dung beetles, manure is quickly claimed and buried by dung beetles. Nitrogen reserves and organic matter of the manure are mixed with the underlying soil, making them available to growing plants and increasing the quality of pasture forage. Excavations of dung beetles increase the permeability of the soil to water and air and clearly improve the structure of soil. In a feed lot, dung just sits there and is pounded by hooves of the confined livestock to the consistency of asphalt, rapidly losing its reserves of nitrogen that escape as gaseous nitrogen in the form of the insidious greenhouse gas nitrous oxide, a gas that has 183-212 times the global warming potential of CO_2.

Dung beetles can also assist as essential allies in enhancing the nutritional quality of the food that we consume. Certain fatty acids, Omega-3 and Omega-6, must be obtained from the food that we eat and are considered essential for health. Dietary experiments have shown that optimal health is associated with a balanced intake of these fatty acids. When the ratio of Omega-6 to Omega-3 exceeds 4:1, people experience more health problems. Our ancestors ate diets in which these two essential fatty acids were balanced, but most modern diets contain ratios for these two essential fatty acids between 25:1 and 20:1. For optimal health and for reduction of cancer, heart disease, arthritis, obesity, diabetes, and autoimmune ailments, people should eliminate the striking imbalance of these two essential fatty acids in their diet. The meat from grain-fed animals raised on a factory farm has a ratio of Omega-6 to Omega-3 essential fatty acids of greater than 20:1, whereas meat from grass-fed beef has a ratio of about 2:1. We practice factory farming in the absence of dung beetles to the detriment of our health. The free services of dung beetles enhance our health by promoting meat production on fertile pastures, by decreasing emissions of methane from livestock raised on high quality forage, and by suppressing escape of the greenhouse gas nitrous oxide from the estimated 829,474 acres of ground that are continuously covered by cattle dung each year.

Modern agriculture with its industrialized management techniques imposes great physiological stress on livestock. The stress imposed on confined factory herds actually promotes the production and transmission of parasites and pathogens. A return to more humane farming practices accompanied by a judicious use of the waste disposal services of dung beetles would also help improve livestock health and eliminate the use of pesticides, antibiotics, and synthetic hormones.

The pages of this book convey the clear message that these beetles of the soil, symbols of regeneration and renewal, should be recruited as partners in our farming enterprises and in our efforts to improve our environment and our health. Industrial agriculturists in their narrow, anthropocentric vision have conveniently ignored the clear lessons of the scientists and farmers who have provided an accurate accounting of the financial and environmental costs of pursuing increased agricultural production with only consideration of its short-term profits.

The hard-won lessons presented in this book are meant to encourage readers to learn more about working in partnership with creatures of the soil. As antidotes to the myopic vision of industrial agriculture, many formal and informal courses, books, journals, and other publications encourage readers to understand and appreciate that dung beetles and their fellow soil creatures are the unsung heroes that, by steadfastly maintaining the balance between processes of growth and processes of decay, improve our soils, our health, and our environment.

Jim Nardi, Ph.D.

Department of Entomology
College of Liberal Arts and Sciences
University of Illinois, Urbana-Champagne

Sisyphus spinipes *rolling dung, a CSIRO presentation redrawn. Named after the Greek myth of Sisyphus, who rolled a stone uphill only to lose control and see it plunge to the valley, the sentence being to roll it uphill again and again.*

Prologue

"A camel is a smoother ride than a horse." I made up my mind to add that line to my notes as I glided along on a Bactrian camel while most of my associates took their pounding on ever-jolting horses. We left the Great Pyramid of Giza on a day trip from the Pyramid of Cheops to el-Sir (pronounced sigh-ear). The camels often did not keep pace with the horses. This enabled a personal discovery that has not entirely evaporated during the intervening quarter of a century.

It was a sandy trail, this ride along the Nile. Animals fed in the evening usually discarded their used feed along the trail, which was free of vegetable growth. Horse biscuits dropped only moments earlier were already being worked on by the time I came along. Incredibly, some beetles were rolling the fresh deposits across the sand, seemingly coating the purloined dung with flecks of sand that caught the sun like so much mica.

Where did they come from, these beetles? This was real desert, not the arid land we Americans call desert in spite of flowers, cacti, brush, and grasses with roots tucked under rocks. This desert drifted with the wind, scoured its foundation as if to desiccate the earth below ever deeper. The cycles that turned the Sahara from a grassland savannah into a centuries-long desert required only 300 years. Those same forces made Australia what it is, a drought-cycle-dogged land forever at the long range mercy of the perihelion, when the Earth is closest to the sun, and the epihelion, when the Earth is farthest from the sun. Add to the above the positions of the largest planet, Mercury, and Earth's neighbor, Mars, plus the Chandler wobble at the North Pole, and you have a good example of cause atop cause until Australia arrives at its six-year drought cycle, a short-term hard times, and finally cessation of the most imaginative event since ancient seekers first domesticated wild animals.

I caught up with Professor Phil Callahan, entomologist, philosopher, physicist, and science writer. "How do these tumble bugs know that there's fresh dung to be had out in the middle of the sand?" It seemed a reasonable question. It certainly couldn't be odor that communicated the message, not in that wind. And there didn't seem to be any bugs on the wing, like scouting planes hunting a polar bear or even an armadillo on a Texas gravel road.

Callahan had just released his magnum opus, *Tuning in to Nature*. It was his thesis that insects communicate in the infrared end of the spectrum, and that they receive emanations in the same way. Yes, the dung sends out a message in the infrared, and antenna-toting insects pick up the signal. Callahan said, "That's your scarab. If you had the scarab's power in terms of its size, given in terms of yours, you'd be Superman."

Callahan's view is not the conventional view. Entomologists by the hundreds have made library shelves sway-back with tomes that turn dung beetles into odor-loving insects that slurp dung juice because it smells good to all the tribes just as most foods smell good to most human beings.

Scarabaeidae sacer became an Egyptian obsession because the beetle united no less than three elements of their culture, sun, soil, cattle. A very large beetle, this sacer character preferred cattle dung to camel droppings. It reminded of the civilization along the Nile, of the sun in its morphology, the rising sun, that is.

The dung ball of this roller bug is shaped in the morning. Its rolling capacity depends on the terrain in a land with a paucity of forage. It travels, this miniature ship of the desert, the way the sun navigates the sky.

The scarab sees to the birth of its young in underground chambers, the destiny of its brood balls. The sun resurrects itself, promising new life. It is said that the god Osiris arose from the beetle's pupa. Nesting, like sex, preserves the species.

The dung beetle Scarabaeidae family seems as important to the mythology of Egypt as is Osiris. It was and remains the sacred scarab. It is likely that Egypt's interest in scarabs was theological. Several species, including the giants of the dung beetles, make brood balls as large as baseballs, some being lodged eight feet deep in the soil. The archaeologists who first found them thought they were cannonballs.

Egyptologists tell us that the faithful considered the scarab the embodiment of the sun god. Did not the sun cross the sky much as the scarab crosses bare ground? The name of the god-like beetle was derived from what may loosely and liberally be translated as "come into existence." It was the embodiment of God the Creator who invented Himself, much like Darwin's life created itself. The business of a beetle emerging from a brood ball must have qualified as self-creation.

The ancients believed that the scarab ilk had no females. It was the injection of sperm into a ball of material that enabled life, all this from a sphere of dung molded into shape with the hind legs of a divine bug rolling the ball east to west, always looking to the east. This fount of life is buried for 28 days. On the 29th day, a water release frees the cargo.

The ancients took note of the battles over the possession of a dung ball. This they regarded as a symbol of courage among the combatants. Roman writers put sheep dung into the equation, and they described what we now call tunnelers. Gods tend to procreate in mythology, and so the sun god led to a god of the rising sun, a scarab-headed man. Thus, Osiris and Isis came into being.

Osiris endured death annually, but he also extinguished death by his annual resurrection. He was the personification of the vitality and self-renewal of Nature. Could those ancients do less than assign the same immortality to the scarab? The *Book of the Dead* has a pecking order for death and funerals. The descendant, the creator god, was murdered by his brother Seth and brought back from the dead by his sister and future wife Isis to be King of the Netherworld. Their son, Horus, the falcon-headed god, brought comeuppance to the murderer of his father, and then ruled Upper and Lower Egypt. Power survived even death, for which reason the dung beetle was empowered.

We are required to correct and update the ancient scribes who have deeded to us their all. Yet, in spirit, Horus, with power from Osiris, is still with us. Those ancients believed in Nature's balance and, not least, in fecundity. The power to vanquish enemies who would destroy good order is now seated in government, the quid pro quo of the social contract.

The beetles of the world constitute the greatest animal numbers. Kings, presidents, and generals all fade away, but the dung beetle, cloaked in anonymity for all but the professional dung beetle watcher, abides.

"Where did the scarab come from and where did it go?" Back from Egypt, Phil Callahan supplied more information on tumble bugs, and he stayed on as a consultant to *Acres U.S.A.* There were so many corners of Nature that required exploration that dung beetles simply dropped out of sight with or without a paean to designate the event. A simple stone model of the scarab, fetched from a Cairo curio shop or a tourist trap near the Aswan Dam, I don't remember which, held down papers on my editorial desk, but it vanished when the office was moved.

The Egyptians were not alone in worshipping beetles, albeit not necessarily dung beetles. To start with, there's the business of Greek language, religion, and folklore being embodied in beetle nomenclature. There's the term Cantharus, technically a vessel with handles, which is a genus in the soldier beetle family Cantharidae. This term once described dung beetles that were believed to find their conduit to life via their rectum. Ethnologists have had a field day tracing language, but I elect not to go there in this narrative.

Legend and myth are more inviting pursuits, and those of us who have dabbled in such will recall the Aristophanes play in which a dung beetle figures in ending the Peloponnesian War. The chief character wonders aloud whether the war can be stopped. He flies on the back of a dung beetle to counsel with Zeus.

J.F. Smithcors, the author of *Evolution of the Veterinary Art,* tells us about early farriers who believed that beetles in pasture grass were ingested as the bovine rolled its tongue across its only row of teeth seated on the lower jaw. This intake, they believed, caused the animal to balloon its belly and explode. Later, much later, researchers found that it was the blister beetle that caused this misery.

Ideas afloat in the Middle Ages are still with us today, Smithcors claimed, notably depopulation when a single bovine tuberculosis reactor is found in a herd, except that the semi-ancients killed only debilitated animals, not healthy ones in the general environment. Children today are sometimes told about the friendly bug named for the Virgin Mary. Our Lady Bug, now just "lady bug," has graduated into nursery rhymes, however the lady bug is not a dung crawler.

Withal, it was Greek culture that made dung beetle lore a staple in stage plays that permitted lots of scatological humor. Down track a ways, Dante's *Divine Comedy* perfectly portrayed the theology of the day, and is said to have coined the word "shit." In today's discourse, it would be the "S" word, a device invented by F. Lee Bailey during the O.J. Simpson trial. In the era of St. Thomas Aquinas, the dung beetle was believed to be the alter ego of the tainted sinner. The business of rolling dung came to mean foul deeds and debauchery. The step to making the scarab a symbol of bad luck followed. My forebears from Bavaria called it the witch beetle, and certainly associated it with the devil, albeit with no particular one. On the other hand, rescuing an upturned beetle by setting it aright again might save a crop from pestilence or a home from storm and fire. Some of that lore made it to Western Kansas where it was bad luck among some Volga Germans and Hungarians to kill the little skinless hard-shelled critter. This devil connection was no more fanciful than the Egyptian's worship of Osiris, Isis, and the scarab. Also called the coffin beetle, the scarab was supposedly a colleague of Beelzebub, who was assigned the chore of consuming the bodies of sinners.

These few facts may be useful if the reader gets on a quiz show, but they are hardly conversation material outside of Ireland, Britain, and Europe.

As we meet a few of the thousands of species, we will get to know them for what they are and perhaps solve some of the riddles that attend their propagation, incubation, and colonization. Some few years had to intervene before I met George Truman Fincher at a San Antonio seminar. Truman opened his lecture with a slide blown up to approximately the size of a small billboard. It was a gigantic "cow platter," as we called it during my Kansas High Plains youth. Truman Fincher had his laconic comment. "This may look like a cow pie to you, but to me, it's my bread and butter."

In fact, it wasn't really his bread and butter any longer. The word had come down from the United States Department of Agriculture's ARS (Agriculture Research Service) to destroy the dung beetles that he'd been propagating, rearing, and colonizing for cattle producers in Texas and parts of the South. You'll meet Truman in these pages, as well as Walt Davis and some few of the human actors in the dung beetle drama.

Most of the papers used to background this story deal with the kind of things that command the attention of entomologists. The details are so clinical that one sees them as extrapolated from that ancient British text, *Grey's Anatomy* (1858). There are excellent manuals that furnish these details replete with Latinate names, genus, species, and orders, most of which can't claim U.S. citizenship, *jus* soil or *jus* blood. Ninety species represent dung beetles, not all of which are members of the family Scarabaeidae.

When a subject has so much literature that it could fill a large Carnegie library, it is always risky to generalize, but to the best of my knowledge, Truman Fincher and his associates are almost alone in computing the economic value of dung beetles. Walt Davis has assured me that his Texas cattle operation never tasted real solvency until he colonized dung beetles and was required by that act to reject toxic genetic chemicals for his ranch.

Most of my readers know that agriculture in general was made a public utility. The agency of that transformation was 7 U.S.C. 601-602, the original Roosevelt-era parity bill. Under its provision, producers of basic storable commodities from 82% of the harvested acres were to be treated the same as labor, business, and even politicians. It was assumed that producers of hogs, cattle, goats, fruits, nuts, and vegetables would rise with the supported commodities the way a rowboat rises with the ocean liner when the tide comes in. Everyone knows that this scenario took a bad turn with the Aiken Bill in 1948. Parity has been scaled downward by Congress via its five-year-interval Farm Bills until now the storable commodity producer receives hardly 20% of the income based on the parity formula.

I recite this in order to make the point that absconders failed to consider the gold in grass called animal manure. All the farmer has to do is treat those unpaid workers well. This treasure trove is detailed in Chapter 13. The bottom line can be governed by the down under denizens we call dung beetles. Let's meet these fellows in all their diversity. Let's watch them turn a disposal problem into hard cash, as if they've discovered how to turn base metals into gold. Those who do not resonate with the lessons offered here might as well compute the values they are missing and add them to the cost of herbicides, insecticides, fungicides, and other chemicals of organic synthesis.

In reviewing a great deal of literature on the dung beetle subject, it seems to me that a preponderance of work has been dedicated to the anatomy of the insect, and less to the way it can contribute to the cowman's bottom line. Indeed, many schoolmen seem to study Nature for its esoteric minutiae and for the ability to communicate with insects, with enough Latinate names to baffle any rancher and destroy the cadence of any sentence. The roster of genus and species runs into the thousands.

In contrast, weed manuals have an easy time of it reaching the layman. For instance, *Astragalus mollisimus* is simply woolly loco, or loco weed. The common names of almost all weeds are colorful, Japanese brome-grass, chess, Oregon grape, broadleaf signal grass, et cetera. Not so with the families, genera, and species of those that savor fresh dung. Almost all are simply "dung beetles." Only rarely do farmers turn a Linnaean Latinate designation into a word for daily conversation, as they do with a Texas favorite, "gazella" from *Onthophagus gazella*.

Dung beetles ask for the positive manager's attention. They are of the order Coleoptera, and belong to the family Scarabaeidae. Zoologists figure that there are more than 90 species in the United States, but hardly more than a dozen are the workhorses of the local tribe. Their reason for being, as far as the cowman is concerned, is to bury dung, to interdict the proliferation of face flies and the equally sinister horn fly. They perform this service with ritualistic dung burial, out of sight, usually out of mind.

But now the plot thickens. Those "cow platters" that youngsters of an earlier era assiduously avoided while playing pasture baseball contain values often left uncalculated. The nitrogen lost by inattention to this unpaid worker turns the bottom line from figurative red to black. The phosphorus available from buried dung makes the paltry uptake generally achieved from NPK fertilizers pale into insignificance. Factories attempt to make those salt fertilizers soluble, but Nature says, "No!" Microbial workers assisted by dung beetles best hold phosphorus in a soluble form for cafeteria-style uptake.

A very astute observer, less an oracle than an analyst, once predicted that the next civilization-shaking invention will be accurate accounting. If so, it will discern the values delivered by Nature's denizens of down under. Chapter 1 sets the stage. All that follows is elaboration, albeit elaboration with a purpose. In the last chapters, the lessons contained herein become morphed into appropriate public policy conclusions.

I rediscovered that stone scarab paperweight on the day that I closed down the research phase of this book. I now see why the insect was sacred to the Egyptians. Data developed by the Environmental Protection Agency reveal that animals in confinement-feeding situations produce three times the amount of excreta of the human population. Little of this used animal feed becomes food for dung beetles because the dung is too concentrated, it is desiccated by the sun and trampled brick-hard, and it is also laced with toxicity. Dung in the pasture is another story.

It is this story I sing!

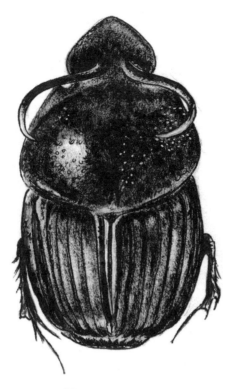

Onthophagus taurus *came out of Europe in approximately 1970. "It just showed up, illegal or otherwise." Officially, the port of entry was Florida or Georgia. It is rusty dark brown with lots of nearly black texture on the shell, but not much on its wing covers. It ranges well into the northern climes.*

1

The Greatest Invention

The greatest invention has still to be made. It may arrive in century 21, but it may take much longer. However, cowman Walt Davis of Sweetwater, Texas and Bennington, Oklahoma may have speeded up the process when he came to his inevitable epiphany the year the cattle market went south in 1974-75. The drop was so precipitous that it shook tough and mature cattlemen right out of their boots. It got Walt to reaching back to when his grandmother gave him a calf when he was 12 years old. "I borrowed money to buy cows when I was 16," Davis recalls. "It took until 1974 for me to realize that I'd never make a profit doing what I was doing."

The year 1974 was a signal event itself. Rachel Carson's *Silent Spring* was 12 years old. Her indictment called into question the conventional lessons handed out by the universities and Extension. If her lessons about toxic genetic chemicals in all their forms failed to resonate in cow country, U.S.A., the reality of sinking prices and annihilated profits did. President Nixon closed the gold window on August 15, 1971, thereby admitting the technical bankruptcy of the United States. Not many farm commmodity producers understood this, just as few comprehended the national direction when an earlier administration injected some $72 billion into the economy in the late 1950s "to get the country moving again."

The greatest invention was a few hundred country miles in the down-track future, as was Tip O'Neill's dictum, "All politics is local." Well, it isn't local, but politics determines the profitability of a ranch in Texas, Kansas, California, or anywhere else. Don't pay too much attention to that snivelling $72 billion that politics injected into the economy in the late 1950s. Figures mean very little once inflation is ignited. But when inflation runs one way, and revenue achieved by a primary producer runs in the opposite direction, the time has come to ask Nature about her bookkeeping and her standing injunction, "Let there be balance and fecundity."

Walt Davis pondered that equation in 1974. Starting with politicians, wages were going up, and service fees were climbing exponentially, yet agriculture withered. Prices of farm chemicals and machinery were going through the roof, yet prices of fat cattle in the pasture or on the abattoir rail languished. No doubt about it, the equation was out of sync.

It has been said that a profession or a group achieves madness as a crowd. Individuals come to their senses one at a time. For Walt Davis, the process started in 1974. First, his health deteriorated. He had contracted a variant of brucellosis called undulant fever. That was the medical diagnosis. "I believed I was getting chemical poisoning," he explained, as I took notes for this book. "I'm still not sure that wasn't the case." Debilitation often imposes "deep think" time. The experts said that the low market skid was cyclical, but they couldn't explain the cause of the cause, not even after appealing to the sun's orbit and the Milky Way, the phases of the moon, or the meat protein available for the nation's dinner plates.

Never explained to farmers, ranchers, consumers, or academia not privy to political machinations was the role of public policy in setting prices at the tailgate. Looked at from the vantage point of the first decade of the new century, prices in 1944 or 1947 seem incredibly low. But don't be misled by those digital apparitions. Living can be just as sweet at an industrial wage of $5 an hour as at $25 an hour if the managers of public policy maintain a structural balance.

Here, Nature seems to echo, "Yes, there has to be balance." But there can be no balance when cattle are poorly fed, then drenched with toxins to remove parasites. Circa 1974, most farms might as well have hung out a skull-and-crossbones sign at the gate. Seemingly secure behind their peer-review data bases, the intellectual advisors dismissed Nature's balance as folly and folklore. Chemists employed by huge corporations with American addresses filled publications like *Farm Chemical Handbook* with enough proprietary poisons to require a half inch of paper. Farmers and ranchers seemed to abandon the skills of their forefathers in order to be conventional and commercial.

In 1947, a U.S. Senator or a U.S. Congressman received a base salary of $10,000 per annum. A row-crop producer became the beneficiary of commodity prices based on par exchange with every other sector of the economy. That year, Congress gave itself a $2,000 raise. In 1948, the 80th Congress passed legislation to reduce prices for basic storable commodities to as low as 60% of the balance of 1947, the last sound base period since World War II. Every five years, a new farm bill reduces those same governing farm prices still more, the consequence being that a man like Walt Davis couldn't seem to make a profit between age 16 and mature manhood. Clearly, public policy would not assist agriculture, except to assist millions to leave the farm. Had Congress treated itself on a par with the treatment accorded to the vast framework of agriculture, House and Senate members would now be paid approximately $2,300 a year.

Forget political direction, forget public policy, forget over-arching economics! We are told that a medieval monk invented double-entry bookkeeping. Walt invoked those principles as he sought to match costs with income. It was no longer possible to match those production costs to income when the costs confronted faltering parity. "I bought into Texas A&M industrial agriculture," Davis now confesses. "I used their fertility program. I subscribed to their animal health program and their best weed control program. I had just finished graduate work at A&M. I knew all there was to know about ranching except how to make money. It was the 4-H calf all over again."

Indeed, this Texas-Oklahoma cowman knew how to get a 95% calf crop. Production hitting the best available norm was not a problem. Davis reasoned that both the devil and salvation were in the details. The slightest change in the market, as directed by either public policy, imports, or supply and demand, set the operation into the reddest of red ink.

Use of over 100 pounds of actual nitrogen on Bermuda grass painted the pastures green, but it demanded the use of weed spray, chemical nostrums too numerous to list, hormones, and everything the pharmaceutical and biological houses could conjure up and sell. The expenses stacked up like warehouse lumber, but collection at the market was found wanting. It was a case of turning to high school business arithmetic, to that hoped for balance that Nature imposes and man presumes to dispose with superior knowledge.

Balance was also being offended by the resource base. Soil fertility was worsening. The nitrogen mistake enlarged itself into a 100-pound, then a 150-pound, mistake. In addition to meat protein production, Walt Davis grew pecans. "The more we sprayed, the more we had to spray. The downward spiral was killing both the ranch and its proprietor."

Those who tabulated rural deaths in the decade of the 1970s discerned the tragedy of young housewives becoming casualties as a consequence of doing laundry chores. Much like the chicken grower with three children and emphysema mused, "I have to be here for those kids, but I won't be here if I don't get out of this business," Walt Davis faced his own epiphany with alacrity.

His reward came a couple of years later. The arrival of dung beetles at the ranch was fortuitous and serendipitous. They seemed to be Nature's paycheck, a debit entry never envisioned by professors of animal husbandry. All of a sudden, all sorts of bookkeeping values dazzled the mind. Modification of T-accounts suggested themselves, if not for paper, then at least as fodder for the thinking mill.

There was the matter of pasture parasites, diseases, pests, grubs, horn flies, all of which spring like Minerva from the head of Zeus the minute a cow evacuates even once out of her eight or nine times a day. Dung beetles remove and sequester dung from pastures while the season permits, thereby enforcing birth control on all such costly enemies of herd management. This new arrival further told Walt Davis that these winged consumers of manure preserved for soil fertility the organic nitrogen and phosphorus in manure, this rather than allowing them to perish via desiccation after hatching all those deficit parasites that hamper beef production. Moreover, swift removal of cow manure prevents grass from making rank growth, in effect taking valuable grass out of production. The mere existence of dung beetles is a greater guardian of the organic red-meat supply than all the inspection certificates and agencies of verification can account for.

Dung beetles, the cowman told himself, preserve grass, and grass puts Omega-3 and Omega-6 opposite each other in a balanced equals sign. The proof is there to see on the abattoir rail, if not expressed in primitive bookkeeping equations. Reduction of processing costs is a given when higher quality products enter trade channels.

The bovine is a ruminant, but not a ruminant hog. When high-protein feed is used to force milk production or growth, much of the feed goes through the animal without being digested. Undigested or rotten protein has a foul odor. Dung beetles will feed on it, but make few brood balls. For the last 4 or 5 decades, some few cowmen have fed the bovine diatomaceous earth, sometimes called fossil-shell flour. Grass from well-policed pastures not only avoids poor digestion, it serves up a manure less offensive to man and animal alike. It is the role of fossil-shell flour to clear out parasites before the cowpat hits the turf. The diatoms in the flour separate feed particles, making it possible to slide over each one for enzyme attack, for complete digestion, and for mitigating the unpleasant odor of unnatural manure. The action of diatoms is both mechanical and physical. In the hundreds of articles consulted in writing this book, I have discovered solid confirmation that diatoms processed by the bovine fermentation vat do not stay on to assault dung beetles when the pat hits the grass. Perhaps academia has not asked the question before. However, the same question regarding the lethal effect of wormers like Phosnet pours to control grubs, face flies, and horn flies is answered by a paucity of dung beetles even after colonization.

Whatever the quality of the manure, its expeditious removal complies with mandates imposed by human health considerations, pathogens being common to both human beings and animals. Face flies hatch in pastures where no clean-up crew survives. They have their marching orders for as long as weather allows them to endure. They live on tiny dung tombstones untouched by feeders banished from their cafeteria by chemicals that many scientists believe have 0-safety level and 0-tolerance level.

Once Walt Davis's new arrivals kicked open the door, there was no closing it again. Here were beetles that performed a service, asked for no pay, and by their presence told a credentialed cowman that it was "back-to-school" time. He started to "do without," in his words. The first order of business was improvement of grazing management. He'd been rotating pastures since the 1950s, this with 3-4 paddocks per herd. In compliance with lessons offered by Holistic Resources Management, with Allan Savory as schoolmaster, with more paddocks per herd, Davis could do a better job as a grazier and follow more closely what cattle were consuming. This management shift enabled him to bring in forage legumes.

At Davis' location east of the isohyet line, rainfall averages 42 inches a year, just right for rotational grazing and dung beetle proliferation. The ranch that became the focal point for dung beetle scholarship is located due south of Bennington on the Red River between the Blue and Red Rivers. The dung beetles had a reason for being in that area. They had been colonized much the same way that Jonathan Chapman once planted apple trees in the Ohio Valley. The "take" had been successful, according to the absence or presence of toxicity.

Around 1986, dung beetles began to show up in quantities. Like a wide-eyed youngster in grade school, the sun-tanned wire of a cowman asked, "Where do they come from? Where do they go?" Looking back, Davis now recalls, "I can explain it. I'd gotten stock density up and grazing periods down to a point where one-day moves were the norm during spring and early summer." These were the conditions that allowed gazella to flourish. Gazella is the last half of the scientific name for the genus and species *Onthophagus gazella*. It owed its existence to George Truman Fincher, a College Station specialist who was busy those days importing dung beetle eggs, incubating and colonizing them in the laboratory, and finally transferring the hatch to the field for real colonization. Truman Fincher is an integral part of this story. His life time of work seems to have been rejected by academia, the United States Department of Agriculture, the powers that be in Washington, and positively pooh-poohed by his chemical nemeses, but not by ranchers who are demanding a return of the values usurped by faceless bureaucrats and second-raters who author internecine strife among scientists. We will meet Truman Fincher again as this story of a balance-sheet value unfolds.

There was in fact an 8-10 year interval between the methodical release and Walt Davis's sudden harvest of plenty. The releases were accomplished in North Texas, southeast Oklahoma, on the ranch of Roy Catullo near Hugo, Oklahoma, west of the isohyet, and in counties and on other ranches too numerous to mention. In almost all cases, management and moisture conditions were critical to the maintenance of a robust population. Not one of Walt's neighbors had anything like the numbers that assisted him to achieve a bottom line.

"Those beetles really got to work," says Davis. "In a paddock just vacated by a herd with 15,000 pounds density, in 48 hours there was no manure! It was gone! Buried deep, and what little the beetles passed over was desiccated by the sun."

A computer and a figure engineer can make the calculations that remand the values expressed to the columnar pad, the income statement, and the dead-reckoning laboratory that lowers costs so that a working profit can be realized.

A full-time, free work force is devoutly to be wished, but Nature has her own stern code. During the daytime, the beetles slow their work. They feed. They seem to know the moisture content of soil that an egg set has to have to survive. Feeding on the moisture in dung more correctly defines the role of the dung beetle. They slurp at Nature's trough, so to speak.

All generalizations are false, including this one. Even after dung beetle activity became an observable fact, Walt Davis came to realize that his fund of knowledge was in deficit. He corresponded with Truman Fincher and became one of a stream of correspondents that trailed Truman Fincher well into retirement. Early on, the lessons were clear. Pesticides and other chemicals of organic synthesis, as well as certain salt fertilizers, destroy dung beetle populations as surely as do fire and perpetual ice. Extremely high stock density, short grazing periods, and 30-day+ recovery periods mean conditions that suit the dung beetle.

The *Onthophagus gazella* has icon status among the millions, even billions, of unpaid workers on any beetle-populated ranch. Davis could count 13 species when fellow ranchers came to his Oklahoma holdings for a seminar. They have names that would please Linnaeus, usually a combination of Latin and indigenous languages combined to sound very scientific.

Wherever they have not been poisoned from the scene, there are natives and escaped natives all over the South. However, modern industrial farming does not permit habitats that enable dung beetles, of whatever species, to proliferate. One of the chest-thumping claims out of the USDA is that the market now has cattle deworming agents that completely sterilize the cowpat before it hits the grass. Left unanswered is the question of what such a procedure does to the meat protein. The bolus method lasts 5-5 1/2 months in cattle and kills flies in dung. Truman Fincher tells me that some of his scientist friends do not care what happens to dung-inhabiting beneficial insects.

For Walt Davis and a few fellow ranchers, dung beetles were the good news, a veritable gospel, but being told that there were tunnelers, rollers, and dwellers merely scratched the surface. Always a student and a teacher, Davis tried valiantly to pass the good news to all who would listen. He invited Truman Fincher to one of his seminars.

Truman, the quintessential good salesman, believed that much instruction is lost unless there is a demonstration. He brought along an inventory of laboratory-reared dung beetles of various species and set up a demonstration suitably caged so that resident beetles could not affect the outcome. Several pounds of fresh dung were supplied for the experiment, and an equal amount was provided for the well-colonized beetles on the farm. During the overnight time allotted, fully half of the dung provided to Fincher's beetles had disappeared. During the same time frame, well-established farm dung beetles had not only demolished their dung ration, they had mopped up the area as well. If there were skeptics in the Oklahoma Holistic Resources Management organization present that day, they could not have had a leg to stand on. The local dung beetles had won, mandibles and power legs at rest.

Unfortunately, gazella rarely become established very far above southeast Oklahoma. They are stern taskmasters when it comes to accepting an environment. It is this realization that has interested cowmen scouting the countryside for *Onthophagus taurus,* the horned beetle that forever reminds aficionados of the fighting bull in the ring. The taurus is a native of southern Europe. Legend has it that this one hitched a ride on equipment from Germany. From Oklahoma northward, this manure-slurping machine is to knowledgeable farmers what desert gravel was to the weary traveler in Russell Conwell's *Acres of Diamonds,* happiness because he'd picked up a handful in the dark, sorrow when the stones turned out to be diamonds, and he'd picked up so few.

Onthophogus taurus has become established along the eastern seaboard, yet hardly one accounting entry makes note of this fact. *Onthophagus taurus* is basically as effective as gazella. The real estate salesmen say, "location, location, location." So does the cowman. Matching dung beetles to environment was the last thing in the minds of the Davis family when they moved from Sweetwater, in retreat from a drought, to Bennington in 1950, hard on the hunt for 35-40 inches of rainfall per annum. No one understood esoteric accounting then, although developments hinted as much.

"We traded one set of problems for another," Walt smiles as he recalls his trail to dung beetle heaven. Rainbelt grass looked awesome, yet cows could starve on it. Fertility was that low, and fertility underfoot was going to waste. It was the tryst with nitrogen that called into question the high-tech of the schoolmen who sounded *mea culpas* about carbon dioxide buildup and global warming.

Fully 80% of the nitrogen in a cowpat out of touch with dung beetles evaporates into the atmosphere. The chemistry of what really happens is as elusive as global warming and the swing of the stars. Carbon dioxide is related to nitrous oxide, and both call into question the hard nitrogens now an integral part of modern agronomy. I will defer to Chapter 19 a fuller explanation of this phenomenon, keeping in mind that real economic value resides in the dung pat, albeit only when it is installed within reach of soil microbes in the soil.

Nitrous oxide, N_2O, is quite similar in molecular weight to carbon dioxide, CO_2. Both are considered greenhouse gases. However, nitrous oxide has 183-212 times the pollution effect of carbon dioxide. Even a gardener knows that nitrous oxide is a decomposition product of nitrogen fertilizer. Anhydrous ammonia produces up to three times more nitrous oxide than urea or ammonium nitrate. One of the most stable molecules is that of nitrogen gas. The oxygen in nitrous oxide is easily lost to produce N_2. The free oxygen atom is then involved in atmospheric chain reactions. Nitrogen dioxide and nitrous oxide are often called NOX, a major "sin of emission" from internal combustion engines. Ozone, of course, is O_3. It easily sheds an oxygen atom in order to clean the air.

The stated reason for no-till farming is that the method apparently releases less carbon dioxide and nitrous oxide, both greenhouse-gas ogres, based on heavy nitrogen fertilization. The service of dung beetles in placing organic nitrogen in escrow can be calculated. The numbers tell us almost as much about inflation for the era as they do about the income statement, except that in any era subtracting nitrogen costs means adding to the bottom line.

In 1962, the farm without dung beetles lost money, and it continued along that ruinous path, based on heavy machinery and nitrogen. By 1972, ten years later, the cost of keeping a mother cow was hopelessly out of line. Sprigging coastal Bermuda, planting small grains for winter consumption, even planting Sudan for summer pasture seemed to create a bottom line that left a sinking feeling in the pit of the stomach. Add some 20 chemicals to the equation and you have the picture. "I can't help you," the doctor said. He was ready to write Walt Davis off.

It may be that the use of such accounting jargon provided the metaphor that gave birth to the yearning for a real, rather than a bogus, discipline, one that calculated the value of the weather when it smiled. Must the markets be written off as far as an individual producer is concerned? If sunshine and rain rate a bookkeeper's attention, why not those billions of microbial workers? Why not the dung beetles that save pasture and grass? Those denizens of the down under can save measurable acres of grass when it is most suitable for grazing. Evolution is no one-way street. Dung beetles and dung-producing animals comply with Nature's balance requirement. Under Nature's rules, it is not possible to damage one member of the community without long-term damage to the entire system. Dung removal and recovery time for plants go together like ham and eggs.

The business of spraying cattle every 21 days contradicts Nature. Spraying pastures with toxicity is equally foolish because cause and effect are mismatched. No reasonable person can make a case for 2,4-D and 2,4,5-T spraying once the debits and credits are sorted out.

It may seem ironic that dwellers in the dark best assist the cowman's harvest of sunshine. There is only one family — Scarabaeidae — and the genera can be counted, but the species are worthy of compendium treatment. They are immigrants from the far corners of the planet, and they declare their specialty, often shunning manures that fail to comply with their genetic codes. Our subjective reasoning is often at fault, for which reason we often arrive at conclusions that do not square with the observed facts of the situation.

That dung beetles are lured by odor to a deposit in the grass or on the sand is usually treated as a given, but this may not be the case. I know of no odor experiments that stand beyond the shadow of doubt. My old mentor and co-author, Phil Callahan, is an entomologist, professor, and physicist. His *Tuning in to Nature* sets out experiments that identify wave lengths in the infrared part of the spectrum as the guiding beacon that brings insects to prey on food the way a corn plant without the hormone dimboa lures the earworm by the signals it emits. Is the cowpat any different? All organic life and residue of life emit infrared wave lengths. To pose this question is to suggest an answer. It may not arrive in so many words, but it is implied in the pages that follow.

Organic carbon has its signal, as do radiomimetic chemicals, meaning chemicals that ape the character of radiation. We can only hint at the complexity of Nature's organization, and we interdict its patterns at a risk that is all our own.

Tumble bugs collect dung and roll the harvest away for burial as hosts for eggs destined to increase and multiply their kind. The minutiae that attend their lives and times make for rich, ripe, even racy reading. Others bury dung in balls that range in size and durability. A third group lives in the cowpat itself, but never spreads the material, and is slow to increase in population.

There was a time when the dung beetle could arrive in the U.S. without papers, usually hitching a ride with cattle, even while the same shipments brought in the horn fly in 1887. The fly spread across the U.S. and Canada because there were not enough dung beetles to control their fecundity.

A cow-calf production facility that becomes a dung beetle haven finds itself with a bankable asset, assuming that bankers understand real assets. The penchant is for the poster-child dung beetle to so derail fly and parasite production via burial that they can't resurface to assert their mischief. Six inches between earth and air is enough to cancel out the unwanted intruders because larvae cannot migrate to the surface to be taken up by livestock.

Why, then, are native dung beetles so ineffective in claiming the American continent as their own? To blame Poison Control Centers, established in 1949, and toxic agriculture for the decline of the natives is perhaps only half the story. Nor is the sweep of the glaciers from north to south in recent geological times entirely at fault. No doubt perpetual ice drove the beetles to well below the Panama isthmus, or killed them, but some beetles have survived, albeit not the most astute and functional of the tribes.

When I was a youngster in western Kansas, you could still see buffalo wallows gouged out of the pancake-flat prairie. Occasional gigantic circles, not greatly different in size from present day irrigation circles, could still be traced in native grass locations. Those circles were as mysterious then as crop circles are now, except that no one speculated about little green men from Mars or some kind of aliens playing tricks on mankind.

I recently encountered an explanation in a book entitled *Frontier Doctor* by Forest Crumbine, M.D. On house calls out of Dodge City, the physician was told that buffalo closed out their grazing day by herding close together, calves on the inside. Bull bison circled the herd all night to protect calves and cows from prairie predators.

Truman Fincher told me that there was more to the story. The American bison had its camp followers. They captured the best of the bison manure even though hooves shredded the droppings as wild animals do in the African Serengeti. When General Sheridan counseled extermination of the bison in order to control the Indian, the resultant public policy also resulted in extermination of the dung beetles that depended on that brand of food. Cows, chiefly Longhorns in Texas and Herefords — and other breeds — replaced the bison, albeit without beetles that favored bovine dung in tow. The shortfall has still to be repaired.

Truman Fincher answered the riddle of dung beetle appearance at the Walt Davis farm, and covered the metes and bounds of beetle population growth or demise. Alabama, Arkansas, Mississippi, Georgia, southeast Oklahoma, and the rest of the rainbelt South came to know his persona and his vehicle. He became a modern counterpart of the soldier-saint Loyola, who once upon a time set out to save Christendom from its folly.

The frontier days of the free range were never very kind to the evolution of native dung beetles. The herds didn't graze and then quit a pasture. More often than not, they overgrazed. Even during a rainy spell, the open range — later, even fenced pasture — did not encourage intense grazing followed by a rest period and a restocking rate in harmony with Nature's balance requirement.

The arrival of political refugee Allan Savory from Rhodesia (now Zimbabwe) called new attention to a movement directed by the Frenchman André Voisin. Voisin's testament entitled *Soil, Grass, and Cancer* provides an excellent exposition and analysis of rotational grazing, a concept that Savory merged with observation of grass on fragile land and installed in the lexicon of Holistic Resource Management. Those lessons dovetailed so neatly with dung beetle colonization and proliferation that they became as one, as the hoped-for resuscitation of a dung beetle ecology got moving circa 1970. Simple cowboy arithmetic and bottom-line accounting demanded it, otherwise — given the public policy outlook on food — the American cowman seemed destined to slide into penury.

The great invention, if it ever arrives, may well acquire a meaningful name, "the Gazella Approach," with the rancher's favorite dung beetle being a metaphor for orders, genera, and species too numerous to have common names and too industrious to strike for higher pay, a perfect example of Nature's order, balance, and effective down-loading of the sun's capital and energy.

Onthophagus gazella *is the workhorse among dung beetles in both Australia and the United States. It was imported from South Africa via Australia in 1972, surviving and propagating successfully in the absence of toxic, genetic chemicals. It has a rusty brown color with charcoal-brown overlay.*

2

A Gift from Africa

Louis and Mary Leakey may well have been within their mark when they assigned the locus for the origin of the human form to Olduvai Gorge in northern Tanzania. The fossils they found in July of 1959 were dated by the then best available method to be 1,750,000 years old. Mary promptly named their find *Zinjanthropus boisei*, genus and species, later changed by other scientists to *Australopithecus boisei*. The original Zin prefix means "out of East Africa."

No one seems to have assigned a similar origin to the dung beetle, in or out of Africa. Did they arrive during the classical Miocene period along with elephants, rhinos, and deer populations with their copious loads of dung? Or was the beetle entry delayed until the Pleistocene period when mastodons roamed the earth some 6-10 million years ago? We know the geological eras the way we know the presidents of the United States, by rote learning, the first by the millions of years embraced by the Paleozoic era, the Mesozoic, on and on until we arrive at the Cenozoic Ice Age, and we encounter worms, microorganisms, and insects. The various life forms all take their place in the pantheon of life, and the dung beetle is no exception. It probably evolved along with the large dung-producing animals, its DNA and chromosomes adjusted through the agency of sex and reproduction.

The explosion of large mammal numbers brought forth a new resource, all in geological good time. The same has been true since cattle started replacing the bison circa the 1870s. Mammalian dung more or less apes the rich humus of the forest, but the post-Ice Age arrival of dung beetles to balance this new resource has been tardy. The rich protein in fresh dung makes it a five-star feast capable of inviting colonies and creating compatible mixtures and an evolutionary return to Nature's preferred balance.

Coleopterists often consider Africa the mecca for beetle diversity. The tropics, of course, outperform every latitude, north or south, to the very edge of those ice-caps not yet extinguished by global warming. Warm climate seems to confer life and color to beetles, especially males, whether they deal in dung or the bark or the canopy of rain-forest trees. Central Africa has been lavish in serving up dung beetles for transplant consideration, yet South Africa seems to emerge as the top provider for programs in Hawaii, Australia, and the American South. Few have scientific names that natives can understand, unless a specific anomaly submerges genus and species by a physical attribute or performance. The elephant beetle has a ponderous horn and a smaller horn below the chief appendage, much like some Errol Flynn-styled coleoptera with both foil and dagger. In the main, Africa's beetles are models by which others are often measured, even described.

It seems that there are no dung beetles that fail to exploit a habitat, assuming the availability of dung. This holds true for the Serengeti, the northern deserts, the top of Kilimanjaro, and the veld of South Africa.

We are content to measure the volume of excreta gifted to the soil by the elephant and wondering aloud at the command performance that dung beetles exhibit in taking it all away. An elephant deposit described by the voluminous exposition of the subject provides food and mass for 48,000 beetles. In two hours the dung is gone. For every kilogram buried, four kilograms of soil are transported to the surface. Surely their DNA directs their activity, for they come off like practicing ecologists, balance-scale in hand. George F. Bornemissza of Australia calculated that 30 pairs of introduced African beetles could bury a sizeable cow pie in two days.

Even South Africa, the home of uncounted beetle treasures, seems dedicated to mega-fauna preservation, and only reluctantly extends endangered species status to its citizens of the dung pile. In the case of the elephant, that pile is significant, for which reason protection has been extended to the favored gifts others have received out of Africa.

There is an African dung beetle that asks for recognition. It is the wing-less scarab, the biggest of its tribe wherever it survives. There was a time when it did its work in eastern Transvaal, Zimbabwe, and other African points, but now survives under protection in South Africa. Fletcher Sims reported back from South Africa that in Kruger Elephant National Park this beetle displays its talent handling elephant and water buffalo dung, but "likes the buffalo dung best," Sims adds. It's a brood-ball builder. But park management significantly managed the beetle's demise by removing for grazing purposes its buffalo-dung providers.

But the elephant is more than a dung machine. It travels far and wide if permitted, often 26 or more miles a day. Naturalists figure the animal marches the equivalent of three times around the earth in a lifetime, and it distributes dung all along the way, albeit not every few steps as does the walking Canada goose. These droppings not only feed dung beetles, but they also transport seeds. Taken deep by tunnelers and tumblebugs, those seeds help provide the lush forage that the traveling elephant consumes on its return elephant walk.

We are told by research giants standing on the shoulders of former giants that there are some 2,000 species of dung beetles on the African con-tinent. They have evolved to fit every climate between the Mediterranean and the Cape of Good Hope. The full roster with even a single line under each genus and species would fill a compendium twice the size of this book. Most ranchers conversant with dung beetles can identify a very few, but almost all know gazella by that common name, or they dazzle their listeners by spilling syllables such as *Onthophagus gazella,* or *Euoniticellus intermedius,* two of the 15 exotics released in time for Walt Davis and friends to discover a shadow economic entry on the elusive cowman's income statement, the term profit and loss having been changed perhaps to eliminate loss from the equation. It may be that some few seekers posted the other 13 beetles so nameless in terms of a common name and so burdened with Latinate prefixes and suffixes. We will meet the cast of characters again with *Onthophagus taurus* and *Onthophagus gazella* being the only two with common names.

Arbitrarily, one has to select the moment in time from which to proceed forward or to look back over the biography of dung beetles. There was a meeting of the nation's dung beetle specialists in New Orleans on September 8 and 9 in 1976. Minutes of the two-day meeting detail how speaker after speaker saw the future of the daring experiment being proposed. The U.S. was to be mapped, then supplied with the correct dung beetles selected from the world's inventory. The economic and ecological benefits were calculated to be so great that rejection of this brave move forward would hardly court opposition. Truman Fincher and his associate, entomologist Richard Blume of Texas, were speakers, but there were others.[1]

Many species of African origin ended up in Hawaii and were reckoned to be suitable for transplanting, first to Australia, then to the continental U.S. because they were more cold-tolerant than others arriving more directly from the Dark Continent. *Onthophagus taurus* became established in Florida, California, and New Jersey quite early, and have since spread up and down the eastern seaboard.

The daring designers expected great overloads of dung as the cattle industry in the South had taken form and was moving straight ahead. They expected more problems, though, one of them being an increase in the use of insecticides as the chemical industry had staked its claim on the land and the animals on it. They must have reasoned, correctly, that insecticides would win the battle as much by annihilating dung beetles as by systemic control.

Fewer horn flies and face flies were expected, even by the promoters of the chemical approach. Fewer problems with intestinal parasites were ceded, as was decreased pasture failure. All the bases were touched with

1 H.L. Barrows, Deputy Administrative Assistant at SWAS in Washington D.C.; W.M. Bruce from the Georgia Coastal Plain Experimental Station in Tifton, Georgia; D.E. Bryan, scientist from the NPS in Beltsville, Maryland; Earl Burnett from the Soil and Water Conservation Research Center in Temple, Texas; R.O. Drummond from the Livestock and Insects Research Center in Kerrville, Texas; R.L. Harris from the Livestock and Insects Research Center in College Station, Texas; Harry Herlich, Chief of the Ruminant Helminthic Diseases Laboratory in Beltsville, Maryland; J.R. Johnston from the Oklahoma-Texas Area Office in College Station, Texas; D.A. Lindquist, scientist from the Insects Affecting Humans and Animals Center in Beltsville, Maryland; W.G. Monson, agronomist from Tifton, Georgia; T.B. Stewart from the Animal Ecology Research Center in Tifton, Georgia; and C.R. Swanson from the Southern Region Office in New Orleans, Louisiana.

suitable refutation, but the dung beetle Tantalus was stabbed and sliced to death with words as pointed as a stiletto blade. Fertilizer nutrients tied up in fecal deposits were to be salvaged, but the documents out of the bureaucracy made light of everything with rhetoric that substituted itself for the insurance man's fine print.

The enchantment with African and other foreign dung beetles notwithstanding, use of native beetles should not be dismissed. Soil types and dung resources are not limited to pastures and bovine animals. When Truman Fincher was a student and station worker at Tifton, Georgia, the discovery was made that the offshore islands held in escrow beetle varieties uniquely suited for swine manure processing. Unanswered from the start was why these native dung beetles were successful on the islands and not on the mainland. This seeming preference for location has consistently dogged dung beetle introduction from Africa and other climes.

During those halcyon days when researchers were contemplating a great American program, the simple premises envisioned earlier enlarged themselves as new questions without answers rose up to bedevil the insightful pioneers. One question that reached out for funding was the location for the daring program. College Station, Texas seemed to have the inside track. The former master of the Senate was now President, and this in all likelihood gave College Station favored status over Tifton, Georgia, the seat of so much research and the locus of interest in tune with a subliminal yearning to discover the "greatest invention" that had still to be invented.

When that select group of scientists met in New Orleans in 1976, the leaders in the field, Truman Fincher included, read their papers, appraised the situation, and concluded that the Australian dung beetle program had failed to validate the claim that buffalo flies and Australian face flies could be brought under control with dung beetles. It was noted that such control of horn flies and face flies had not been achieved in the United States either. Nay-saying was not the order of the day, however. There was considerable data to suggest that intestinal parasites indeed had been slowed, even stopped, in select situations.

Now the variables were rolled out. The open range concept still held on in the West, and sparse stocking rates called into question some of the fondest premises of the dedicated dung beetle promoters. High on the agenda was the matter of a single location for research, ergo money resources out of garden-variety appropriation bills.

A new quarantine facility at College Station made that location the hands-down winner as an importation destination. It was noted that geological development and the monumental switch from native animals to cattle had caused many species of dung beetles to fade from the scene. Evolution requires time that no cowman or practical researcher finds enchanting. Therefore, importation was devoutly to be wished. Tifton, Georgia collaboration aside, sourcing, first from Africa, then from the Americas not scoured by glaciers, then from wherever native fauna and complementary beetle development suggested both fecundity and balance, arose to guide the debate.

Truman Fincher wrote his Ph.D. dissertation based on findings in the Tifton orbit. He drew special attention to the island off the Tifton, Georgia station which was heavily stocked with cattle and swine, as well as with wildlife, and where animal feces achieved complete burial through the agency of indigenous dung beetles. The same situation was noted on Cumberland Island. But not enough was known about the biology and ecology of those islands to make transfer to new and different sites practicable. Finding those answers well away from the rainbelt climate of Georgia was not a reasonable probability either, Fincher held.

I have not burdened these paragraphs with the names of those count-less workers who rate a Hall of Fame all their own, but at least a few side departures should be permitted. The foremost taxonomist of the era in question was R.E. Woodruff of Gainesville, Florida, a collaborator with Truman Fincher on Cumberland Island. The grass breeder, Glenn W. Burton, was also an associate of Fincher during his dung beetle research at Tifton. This triumvirate constituted a powerhouse, but it failed to carry the hour.

Having imported plants, insects, animals, and yes, diseases for several centuries, the United States Department of Agriculture decided to close the door even on this least likely pathway for entry of infectious diseases, opining that dung beetles could be vectors. This has not worried workers from foreign climes who have seemed quite willing to transport College Station eggs back home without the benefit of quarantine.

As bureaucracy and legal interpretation stepped forward to claim dominance over the national program, the potential widened. There was a Pollution Control Act in 1972. Clearly, dung beetle tunnelers could create an open sesame for water pollution control. A flush of urine or rain wash manure could hardly make it to a stream when the soil was laced with tun-nels that turned the ground into a veritable sponge.

It has been said that a science project director has to exhibit a willingness to be bored in order to be successful. But there was nothing boring afloat as the most exciting project since the Lewis and Clark Expedition got under way. At least half of the effort was to further examine that gift from Africa and elsewhere that had already served Hawaii and Australia so well.

The budgets sought were modest enough, a mere $500,000 for Fincher and Blume to take their work forward and for the complex of Ag Research Stations to become involved, all with objectives that must have seemed as ambitious as President Kennedy's determination to put a man on the moon.

Onthophagus gazella, that nonpareil gift from Africa, imported by way of Australia, was more than a poster inspiration. It had become established along with a diversity of associates. At the time, there were 1,137 species of dung beetles that had been described in the Western hemisphere, 900 of which proliferated in South America, 157 in Central America, and 197 in North America; 87 of these last are in the U.S., 40 in the Southwest and 27 in Georgia alone. The discrepancy between the fragmented totals and the total for the hemisphere is due to the fact that some beetles can be found in more than one area. Finding the correct needles in the world's haystack would be an awesome task.

If my reference to a gift from Africa is merely a metaphor, it is probably because no one has attempted a comprehensive genealogy of the world's dung beetles. The African beetles that proved so successful during the heyday of beetle research often came from Africa, especially South Africa. Others tracked their way into Agriculture Research Service compounds from every corner of the world, the assumption being that they were of age where they were, evolving along with the animals that provided their sustenance. If the list is short, we even dare recite it just to prove that citations are not fiction and that someone mastered enough Latin to create their Latinate nomenclature.

In 1986 seven exotic dung beetle species were released in California. When the assignments were handed out, a first order of business was improved rapport with Australia. As if to capture a line from the musical, *Oklahoma,* the New Orleans conference might have said, "The cowman and the pasture beetle should be friends." The project might take dung beetle discovery and propagation into many strange cul-de-sacs along the way, but the clear understanding was that pastures and cow-dung removal come first.

Control of parasites on pasture via natural methods vs. chemicals has divided agriculture ever since chemicals of organic synthesis achieved a scientific imprimatur. The dismal failure of the latter forced all except the most obdurate to revisit what every Depression-era child had seen — dung beetles rolling their treasure from the scene of plenty, disappearing down under and, curiously, doing it backwards. To any youngster growing up in the rural South, and even as far north as Minnesota, a dung beetle was a ball-rolling scarab, and every one of the scarab ilk was the metaphor for all dung beetles.

Figures for 1987 tell us that farm revenue from poultry and livestock was $22.7 billion, more than cash receipts from row crop grains and cotton, this in 10 of the 13 states that comprise the region called southern. With the value of dung beetles a given, farmers came to consider the pasture deficits. They are the house fly, the stable fly, the horn fly and the ubiquitous face fly. Not one is a U.S. native, but the damage is home-spawned, so to speak. Together, the nemesis status would clock in at millions were that longed-for newly invented accounting system a reality. The cost is generally a given, with or without control measures. Without the assistance of the *Farm Chemical Handbook,* the numbers become staggering. With chemical control, a yawning chasm in the checkbook becomes the observed result. The proximate cause has been and remains the dung pat on the pasture. Only the dung beetle propagated in abundance is capable of changing credits into debits.

Hardly a quarter century ago, flies that hatched in the very dung that beetles could have consumed cost farmers and ranchers $100 million each year, and that figure can be factored upward using the President's economic report inflation index. No one has ever attempted to calculate the loss of capital and earnings suffered when health officials actually close down a fly-generating facility. It happens, and not without reason. Flies of several stripes are vectors for typhoid, cholera, amoebic dysentery, anthrax, tuberculosis, conjunctivitis, plague, yaws, even leprosy.

A garden variety pair of flies can propagate another 1.8 million pairs in 6 generations, or 12 weeks, and many have developed resistance to insecticides. The blood-sucking stable fly is not as prolific, but it is an efficient retailer of streptococcus, various bacilli, infectious anemia, tularemia, pinkeye, and swine erysipelas, among others. In her wisdom, Nature has handed these several problems to the miscellany of dung beetles that compete for food when the cowpat hits the grass.

The key to excellent control is competition. Two species, *Onthophagus gazella*, a gift from South Africa, and *Onthophagus taurus*, a gift from Europe, are blood enemies, therefore partners in salvation. Their introduction into Georgia and Florida as early as 1938, 1945, and the 1970s was a prelude to releases elsewhere in the United States. The release of an additional 13 species carried the species and tribes into Alabama, Oklahoma, and Texas. In spite of their lineage, they all answer to the appellation of dung beetle. The African origin, traceable or not, has now become submerged into the grand mosaic of the whole. The sweep was so significant that many dung beetle scientists figured that repopulation of the entire country was imminent.

Releases in the southern U.S. have left a trail as wide as a forest fire. In 1981 the distribution of *Onthophagus gazella* and *Onthophagus taurus* caught the attention of entomologists, especially students, who hastened to orient their college papers in tune with the greatest insect experiment since sterilized releases came close to annihilating the screwworm problem in Texas and the South. By the time cold weather closed down dung beetle activity, the two species had spread from their original release sites in Texas and Florida, crowding their way to Louisiana and Mississippi counties airline miles away. These gifts from Africa answered a cattleman's appeal that was more wished for than expressed in those days. At that time, Truman Fincher computed that dung beetles in a rapid rotation pasture had transformed the area into succulent grass, had abetted the recycling of nutrients, had reduced the helminth parasite and pest fly populations, and had increased yield figures — adjusted for inflation — that spelled the difference between excuses at the coffee klatch and black ink on the bottom line.

Economic calculations didn't cause adrenalin to course through the veins of entomologists as they were busy with the prospect of finishing a national map, identifying dung beetles for each rainfall area and for their ecological suitability for each general climate, the objective being survival season after season and not just during the study tracking season.

Onthophagus gazella, an Afro-Asian beetle was originally released in April of 1972, precisely when natural organiculture was picking up steam and when non-toxic approaches to all phases of food production were asserting themselves. Releases of African immigrant dung beetles were made in 1972, -3, -4, and -5, involving enlarged areas, according to entomologist Richard Blume. The species also became established in

California and in Georgia. It took 3 years for the immigrant beetles to establish themselves in an 8-mile radius.

By the 1990s, the case for dung beetle colonization across the country had been made beyond all refutation. It was then that a new nemesis asserted itself. Truman Fincher at the Animal Parasite Research Lab in Tifton, Georgia used parasitized steers to deliver dung to beetle-populated pastures. Parasite-free animals then grazed the pastures for 45 days. The result could have been expected. Calves on dung beetle pastures picked up markedly less parasites than those on unprotected grass. With cattle stocking rates ever increasing in the Southeast, due to grazing climate and forage, carrying capacity was being cranked up each year as NPK fertilization gathered speed, these recommendations being the current coin according to the farmers' intellectual advisors. Heavy stocking in the absence of a manure removal program put parasite damage on a collision course with insolvency as cowpats canceled out nearly a square yard of pasture per dropping. Gastrointestinal parasites staked their claim as their manna from heaven descended.

Bovine pasture feeding and production of pre-digested food for the dung beetle answers to logic and the observed facts. A good system provides animals with enough feed for the manure production needed by the dung beetle. Cows will usually eat 5 of the 8 inches of forage in a pasture.

They will consume 5 inches of 400 pounds of dry matter per inch on an acre basis in, say, 14 days. The available acres can be divided by 14 to reveal how much forage is available per day. The astute cowman divides per day availability by .03. That tells the cowman how many pounds of animal can be fed. In short, the lactating animal consumes 3% of her body weight per day. A mature or non-growing animal requires, say, 2%. It is body mass that is being fed, and manure production follows accordingly. It is not incorrect to say that the dung beetle gives a good accounting of itself in handling this fecal payload.

Evolution has come to the rescue in many countries as varied dung beetle populations have put parasite depopulation on the other side of the equals sign. The gift from Africa was not only the genera and species that one finds enumerated as this dung beetle saga unfolds. As early as 1967, the intelligence arrived that dung beetles occur in profusion in the excreta of both domestic and wild animals of South Africa, one type supporting the other like two tipsy celebrants in an Irish pub.

Circa 1980, three exotics were released in the U.S., and also were given a new home in Hawaii, the objective being to correct the pasture ecosystem. By now the idea of that great new invention called accounting was taking hold. College Station, Texas entomologist Truman Fincher was computing in terms of hectares, which I usually convert to acres at .405 hectares to the acre. In any case, acres lost to grazing in the United States totaled millions, not mere thousands, due to dung buildup and contamination. Even though dung beetle populations proliferated, they were seldom equal to the task with or without African species on line. The style of grazing held over from open range days was part of the fault. Nature's gift seemed inadequate since bison had been removed from grasslands, replaced by Spanish cattle and Herefords, plus other species, along with horses and mules. A measure of impatience figures in man's assessment of the situation. The young country has not had the time to evolve the climax crop of beetles required by the bovine interlopers as has been the case in Africa.

The late 1980s plan was to install the right beetles with the right voracious appetite required for manure by type. Contamination is in the eye of the beholder. The dung beetle usually likes slurpy fare, not the dry pancake offered by a dessicated pat, whether dessication occurred in the lower colon or in the open air. There are always incidents in the course of human events that drive home lessons otherwise passed over in the rush of life. I am reminded of the story in *Arabian Nights* about Abu Hasan and his pre-marriage feast. Merriment prevailed, and toasts consumed the better part of the evening. At last, Abu Hasan stood up to say a few words. Then, in the words of the narrative, he "let loose a fart loud and terrible." Now, in the Arabian society of the day, it was perfectly correct to belch. In fact, a deep belch was a salute to the food preparers, but sound-supported expulsion accompanied by gas was not acceptable. Abu Hasan left the feast and banished himself for almost two decades, finally returning furtively in time to hear a young girl ask her mother, "Mother, when was I born?"

"Thou wert born the night Abu Hasan farted." To Abu Hasan's utter mortification, his classic faux pas had become one by which his society now dated events.

This came to mind the night I attended a fancy dinner. A bowl of unidentifiable material was served, and called "soup." There was absolutely no moisture in the bowl, and my wife advised me to eat it with a fork. "I feel like a dung beetle," I remarked, "that's been served its dinner by a constipated cow." The dung beetle likes juice. If the pat is suitably moist, an adequate work crew can take it underground in 24 hours.

Thus, the time arrives to compute the tons of nitrogen lost into the atmosphere. An additional computation asks for a number to designate the loss to helminth parasites, all this on the day that the greatest invention of mankind comes to fruition. The benefits that can be chalked up to fly removal can be computed at billions of dollars annually. Assigning these values to each farm requires detailed information that primitive bookkeeping systems are slow to provide.

It is the ecological selection of beetles that returns us to that gift from Africa, which has been the mecca for dung beetle diversity. These ball-rolling scarabs are there by the hundreds, if not thousands, of species. Dung-burying beetles called tunnelers, each specializing in the dungs of Africa's diversity — impalas, zebras, wildebeest, even lions, apes, gorillas — are so specific in appetite that scholars have classified them that way, often embodying their attributes into the Latinate appellations that now fill volumes.

And then there are those dwellers that complete their life cycles within the dung deposits themselves. Perhaps all are Nature's handiwork originally, a gift from that eternally fascinating continent, Africa.

The family Scarabaeidae has some 30,000 species, 1,500 of which reside in North America. In our shooting-from-the-hip style, we say dung beetle actually for three subfamilies of Scarabaeidae. The first two have preference in our lexicon and in our temperate zone latitudes.

This is not to say that all residents of the cowpat except dung beetles are destructive. Many parasitize or prey on face flies and horn flies. In some cases, they compete with those nemeses of the pasture for a food supply. Climate, rainfall, and ambient temperatures also figure in. There have been reports out of Missouri that the state is favorable to a horn fly control development with over 90% mortality in the fresh cowpat. It is probable that the horn fly torments cattle and cattlemen the most. It reportedly came to the U.S. in 1887 through New Jersey, and has since taken its toll not only by sucking blood, but also by disturbing livestock on pasture so that they lose weight or inhibit gains. The cost in late century 20 was estimated at well over $700 million, and by century 21 is certain to exceed the $1 billion mark.

Insecticides, loudly proclaimed in slick farm-paper ads, are marginal control tools at best and worthless as a worst case scenario. Pests, much like weeds, develop resistance to toxic genetic chemicals. The horn fly hones in on a cowpie within minutes, even seconds, after it is deposited. This can happen over 8 times a day per animal. It takes 8-30 days for the egg to hatch, temperature governing the blessed event. Dung beetle activity interdicts the hatch, and therein lies the less than well kept secret of African fertility which has now been released to the world.

The face fly is of a different stripe. It does not suck blood, but it has distressed the comfort of man and animal alike. It came to Nova Scotia as an unwanted gift in 1952 and tripped its way across the temperate part of North America faster than rabbits spread across Australia. The face fly annoys and disturbs. Since cows have no hands to shoo it away, it hangs around the mucous secretions of the eye, often instigating pinkeye. As a consequence, cows stop feeding, bothered as they are by pinkeye and worms. Add the economic loss of well over $50 million yearly plus inflation to the horn fly tab, and the loss becomes a lens through which the benefits of dung beetles achieve focus and size. This mischief cannot be sustained without 15-20 days for the dung-based hatch to become airborne.

A recitation of fly species north of Mexico would disturb the equilibrium of almost any reader except a dyed-in-the-wool aficionado. Richard Blume published the intelligence that, north of Mexico, 457 insects make the cowpat the focal point of their lives. His map shows their distribution. It's a wonder-world that looks more like a universe than a local incident in a remote grassland. Those insects compete, engage in wars, establish Maginot lines, and drain the life out of their enemies, including the horn fly and the face fly, and always obey their DNA advice to increase and multiply. They do not always evolve by mixing XX chromosomes with an available Y, as is usually the case with mammals. They violate every law of genetics and often seem capable of dispatching the hormones needed to prevent either male or female births.

There's a slab of stone atop Mount Everest that once rested at the bottom of the ocean. Its fossil contents tell of life forms no longer on planet Earth, neither in the sea nor on the land. Evolution and the steady shedding of the genetic payload via sex, procreation, and eons of life have so altered life that it takes scientists and conjecture to make the connection. Those who come aboard to make the dung beetle connection have not taken the evolutionist's long and difficult data as gospel. They have reported the story as they have found it, certainly in Africa, but also in almost every other corner of the earth.

This much stated, it should not follow, but does, that parasites perform poorly in holding the face fly and horn fly in check. The burden falls back on man's control of the pests.

Not many cattlemen memorize these names the way a baseball fan memorizes batting averages. A beetle either does the job or it doesn't. It was no accident that three African species gained first attention at College Station, Texas under the tutelage of Richard Blume and Truman Fincher. By the 1980s, hundreds of researchers were hard at the task of unravelling that part of Nature's secrets embodied in dung beetles. Some were dilettantes, interested in getting a grade or busying themselves with esoterica the way a stamp collector fondles stamps with no other purpose in mind. Tagging every sentence, every syllable, every iota of material with a citation makes English text unreadable, or at best, as John Barrymore put it, "You're taking a bath upstairs and someone is ringing the doorbell downstairs."

Onthophagus gazella and *Euoniticellus intermedius* became model species for research as soon as it was discerned that both were unique producers. They seemed to endure the special egg treatment mandated by officials ever on the hunt for pathways for communicable diseases. Africa was suspect to a fault because of exotic syndromes endemic on that continent. Treatments included the stalling down of the importation procedure, but it was not closed down entirely. Opening and immediately closing the egg chamber of the brood cell, replacing the eggs in the original cell, and placing the eggs in distilled water with a detergent solution of 3% formalin outlines the procedure. *Onthophagus gazella* eggs survived the trauma with a superb hatch as the end result. Just the same, the survival of the species was reduced markedly, the price tag for questionable caution. It was the formalin bath that caused the greatest mortality. But it had to be. Bureaucracy wouldn't have it any other way. Foreign visitors to College Station, Texas merely shrugged their shoulders and accepted sample batches of gazella and taurus, secure in the knowledge that the trip from Africa by whatever route would be impeded by similarly strict laws back home.

Importation of exotic dung beetles was started in 1969, control of the horn fly being the objective. Dung beetles and horn flies compete for the same delicacy, this according to instructions installed in their genes and DNA. Introduction of livestock diseases via dung-slurping beetles is a human concept. Dung beetles would hoot at the idea. How could their orders from Nature make them the vector for rinderpest, sleeping sickness, malaria, and a host of tropical diseases when Creation's evolution had decreed otherwise?

This cleansing ritual according to the regulations of the USDA and the Animal and Plant Health Inspection Service mandates that beetles from Africa or elsewhere be bred in culture in the country of origin, as described above, this if a livestock disease is a matter of record in Africa, albeit not in the United States. Quarantine takes over when the gift from abroad arrives.

Importation of foreign dung beetles fleshed out every vestige of paranoia. The beetles were reported to be infested with mites, nematodes, fungi, bacteria. They might be carriers of rinderpest or hoof-and-mouth disease, renamed foot-and-mouth when it was learned that the keratinous pads on mice might also suffer the disease. Usual sanitation treatments were deemed unsatisfactory, for which reason an overkill regimen became the norm. Thus, importation required removal of eggs from dung balls, washing, container entombment, and transport under quarantine. At the destination, the eggs were inserted into artificial dung balls created from now domestic dung. The hatch and Nature's routine followed until a new generation could lay eggs. These eggs then underwent sterilization. Release from quarantine enabled field colonization. That's how *Onthophagus gazella* became a household word among coleopterists in Australia and the United States.

The old sourdough in *Paint Your Wagon* may have been more correct than his flippant line suggested, "Of course there's gold in them thar hills, elsewise where would the idea come from?" Where did the idea of repopulating a nation's pastures with exotic imports come from? As the program and some of the results got under way, all roads led to Australia.

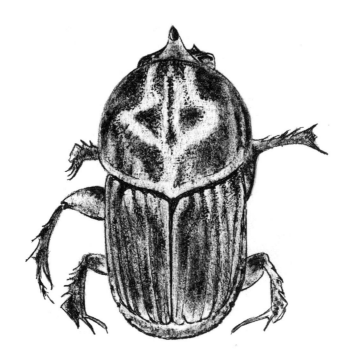

Euoniticellus intermedius, *an import from South Africa via Australia in 1979, is almost an orangey, light rust color with a darker pattern on its shell.*

3
From Down Under

The idea of calling on immigrant dung beetles for pasture improvement was born, not without umbilical cord, in the mind of George F. Bornemissza, who made his suggestion a matter of record in 1960 in the Australian Institute of Agricultural Science publication.

Bornemissza came by his insight quite naturally. Born in a low-tech, highly nature-oriented society with an ancient culture of husbandry and sustainable agriculture, he observed and absorbed the lessons Nature had to offer, one being that animal dung disappeared almost as fast as it was deposited. He noted that certain dung beetles had preferences, a sort of symbiosis between animal and beetle via the agency of dung. When he arrived in Australia after World War II, the new country seemed to ask for the wisdom of the Hungarian peasant.

By the time the Americans consulted with Australia, the objectives of pasture management were well in tow. It was understood that pasture eco-systems depended on the "normal functioning of its nutrient cycle." The words are those of George Bornemissza, who reported on the Australian experience as early as 1951. Then, as later, he made the point that pastures are more than stands of forage grown on land too poor for row crops. The components are many and, should any fall out of sync, the repercussions are awesome.

I could hardly help noticing one of them on the day that I and Harvey Lisle, the author of *Enlivened Rock Powders,* followed a trail of face flies through half of New South Wales to reach the Alex Podolinsky farm. Here was a farm as ecologically sound, it seemed, as possible. Yet face flies introduced both of us to the Australian salute, a constant waving of hands to shoo away the pesky tormentors. Podolinsky's cows consumed grass. His pastures were well treated with biodynamic preps, which he made for farmers handling three quarters of a million acres. Without excess protein, feeds were well digested and cannot be said to emit really foul odors.

Bornemissza could have supplied the answer at once. It was the absence of dung beetles that failed to state the case via the accounting system and even the best of farmers couldn't repair the deficit. Truman Fincher reported to American readers of scientific literature as follows: "The idea of improving pastures by introducing species of dung beetles from other countries was first proposed by Bornemissza in Australia. In fact, the Australians have successfully established several species of dung beetles from Africa in their country (circa 1970s)."

The objectives were the ones discussed at that New Orleans meeting in 1976. The reasons for being of the Australian program were essentially the same as those that became evident in the U.S. The Down Under grassland eco-systems were profoundly disturbed, not by a falling meteorite or a moving glacier, as may have been the U.S. case, but in fact by the arrival of domestic livestock, as was the case when bison were replaced by cattle, sheep, and horses in the U.S. In terms of time, the change was cataclysmic, but dung beetles are slow to adjust to strange new meals.

Prior to the arrival of prisoners on "that distant shore," that continent's herbivores and marsupials probably never delivered enough dung to the eco-system to be a problem. Apparently, such dung as there would have been was easily handled by native dung beetles. This handling speeded decomposition and inserted nutrients — nitrogen, phosphorus, and unnumbered traces — back into the soil. The Australian drill of a platoon of dung beetles pouncing on a dung pellet and tearing it apart in preparation for interment would make grand entertainment if captured by a slow-motion camera.

It was the introduction of domestic grazing animals that upset a cycling system that had obeyed Nature's demand for balance and fecundity. Horses, cattle, and sheep overloaded the system. The mere volume must have been to natives what the hordes of Genghis Khan were to a civilized Europe. Suddenly, valued grasslands were fouled. Those pesky face flies mentioned earlier had too little freezing weather and virtually no bats or birds capable of keeping up with the traffic.

Australia, then as now, suffers the curse of the open range. Dung was scattered far and wide. Even so, native dung beetles and other coprophagous organisms more or less gave a good account of their office. Then came the post-World War II push. High test fertilizers made the scene and with them came chemicals of organic synthesis.

Higher forage yields increased the carrying capacity of each acre or hectare, this without an increase in dung beetle capacity. The business of cowpats remaining undisturbed for months gave rise to all sorts of parasites in addition to veritable swarms of buffalo flies and face flies. Bornemissza came forward. There were dung beetles capable of handling the traffic. There were species in Africa that could handle a washtub half-full of elephant dung. There were the gazella relatives already at work in Hawaii. Agriculturists need to be told that insecticides and herbicides cancel out the ability of Nature's unpaid workers to even survive, much less set new records for dung disposal. Moreover, using medicinal poisons systemically meant beetle annihilation because no beetle could survive being treated like Borgia's guest.

This last realization came down on the fondest expectations of dung beetle promoters like sledge-hammer blows. Field experiments proved in southern Africa and Australia finally revealed for all to see that dung beetles and insect predators could cause 98% fatality of dung-breeding flies. There is at least one chemical that can kill fly larvae and leave dung beetles undisturbed, but the fact is that poisons are poisons, and they have side effects even if they do allow a scarab to make it to the burrow or a human being to the door.

Beetles can also fall prey to vertebrates, especially if beetles exist in large numbers. In African savannas omnivorous mammals feed on the great concentrations of the largest dung beetles, but this did not prove to be a problem in Australia. Suffice it to say, where dung beetle species dominate, the other members of the dung community falter and seem insignificant.

In spite of the varied successes of the Australian dung beetle program, "our authorities don't even want to know about them," Australian John Feehan told me in an exchange that went on for an hour. Feehan had served the agency that had been assigned the dung beetle project at the start. The locus of dung beetle research and repopulation was the Commonwealth Scientific and Industrial Research Organization (CSIRO), which received its funding from farm organizations and the Australian Treasury Department. The matching funds arrangement was dollar for dollar. True to form, the large farm organizations exhibited little interest in dung beetles but innovative farmers were not only interested, they were "gung-ho to get with it." There wasn't a dairy producer in Australia who wasn't aware of beetles that matched his climate, his weather, and his soil. Australia had its own nay-sayers, such as the U.S. found expressed in Wisconsin Senator William Proxmire's "Golden Fleece Award." The general population is understandably not tuned in to insect research and hardly comprehends the economic role of the dung beetle. With an area as large as the United States, Australia has droughts that line up with the late great U.S. dustbowl for comparison. Most urbanites probably saw the dung beetle program as a boondoggle of gigantic proportions.

Still, there were reasons for supporting the program that might well be considered Australia's own. Beetles are capable of producing millions, even billions, of tiny holes in pastures that can trap herbicides, insecticides, wetting agents, fertilizers, organic matter, all of which are killing rivers with runoff. Paddocks with their teeming populations of dung beetles develop so many holes and tunnels that runoff to nearby streams becomes impossible. The micro-bacterial activity thus engendered simply neutralizes the pollutants and even makes them whole again.

The CSIRO devoted 31 years to dung beetles, in effect pioneering the modern attempt to find beetles that would match the environment. For perhaps the first time, here was a major scientific institution calling attention to the flaws in the accounting system that permitted a natural economic plus to be eclipsed by costly industrial outputs, ignoring the phosphorus, nitrogen, and trace minerals that the bovine gifted the producer. The only requirement was cooperation with Nature's helper in invoking that greatest invention still being developed.

Spreading the species imported from Africa by way of long colonization in Hawaii to their climatic and geographic limits, and mapping the location of the same, became workaday stuff as the program expressed itself. The CSIRO dung beetle project has still another milestone to its credit. Chemical firms took notice. The theory was afloat that chemicals dumped on pastures could hammer into submission the parasites, the buffalo flies, the face flies, all of the deficits perceived to reside in the dung of cattle. Only Fort Dodge Laboratories has a preparation that still gives dung beetles permission for life. Even so, chemicals of organic synthesis are not friendly to cattle or grass.

Feehan explained it this way, "If you and I sat down and had a beer at the end of the day, we'd enjoy drinking it. We'd have a good conversation. Within 24 hours our livers would have removed the impurities from our bodies. Some of Merck's drenches are very much like this. They do the job. They kill the internal parasites, and the animal is clean for a few days, but it can be recontaminated with internal parasites within a week. Fort Dodge's product is different. It bonds with the fat in the animal. Because of this, it gets relief for a few days. You could say that the product uses a program already built into the animal. The molecule does not harm dung beetles."

A total of 44 species was released, each keyed to its respective climate. Those species had been harvested from the far corners of planet Earth. Australia shared its treasure with other interested nations, the United States included. Then the funding stopped. Perhaps the chemical firms, never friendly to the repopulation idea, made their clout felt. Perhaps the end came when drought arrived and farm groups withdrew their participating funds. Australian weather is cyclic, as Aussie meteorologist Alexander Gaddes explains in his disarmingly simple *Tomorrow's Weather*. These cycles are governed by the swing of the planets, if not the stars, and by the earth's perihelion and aphelion, or closest and farthest distance from the sun.

We won't go there in this book, but we have to exhibit some empathy for the farmers who no longer had the money resources to continue the Australian dung beetle program. Government economy measures are often a lot like an exquisitely stupid robbery in which the thief steals the goods with the least value. The entire cost was hardly more than the expense of a Prime Minister's photo-op trip.

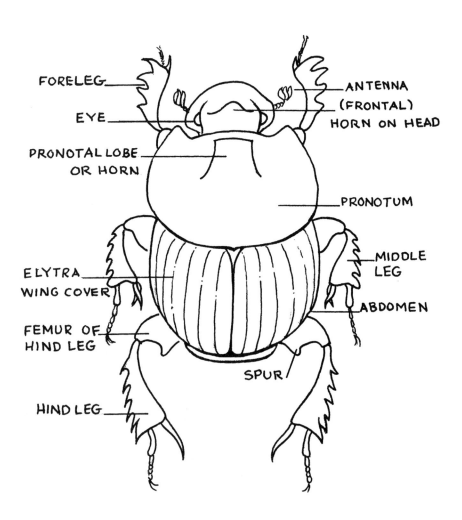

FORELEG

EYE

PRONOTAL LOBE
OR HORN

ELYTRA
WING COVER

FEMUR OF
HIND LEG

HIND LEG

ANTENNA
(FRONTAL)

HORN ON HEAD

PRONOTUM

MIDDLE
LEG

ABDOMEN

SPUR

The experiences and lessons were not lost. The dung beetle specialists, some of whom you've met in Chapter 2, added species and insight to their larder. CSIRO published a profusely illustrated little book entitled *Common Dung Beetles in Pastures of South-eastern Australia* by Marina Tyndale-Biscoe, but it too has passed from the scene and is only rarely available in used book stores. The hoped-for consequence of the program's cessation, in the eyes of some disinterested groups, was endangered, and "vanished species" status for the dung beetles was installed on "that distant shore."

CSIRO legacy has been to pass along almost encyclopedic knowledge, including a manual for identification of imported species as well as native fauna. We can start with our metaphorical gazella and refer readers to the gamut of articles and manuals that make library shelves founts of knowledge that only tabulation rosters and data bases can accommodate. The nomenclature of the atypical dung beetle complies with a Down Under diagram circulated by CSIRO.

A diagram tells less than half the story. The hard outside cuticle is a given. Three pairs of legs may be one of Nature's anomalies, but they are standard. Also standard are barbs on the forelegs of the dung beetle, which makes it difficult to dislodge them from flesh or fecal matter. Folded wings protected by an encasement except during flight are a mark of dung beetles. The repetition of the words "dung beetle" may seem redundant until the reader is reminded that there are other beetles that feed on plants, but do not participate in dung removal.

Many beetles are very handsome bugs, often possessed of bright colors. The division is clear. Natives that evolved with wallabies, kangaroos, even Tasmanian devils, seemingly could not make the transition from pellets to cow platters, and were left to fend for themselves, while dung beetles from the proverbial four corners were coaxed to stake their claim.

The rest of the Australian story becomes an entry in the grand mosaic of the whole, as the baton is passed to one worker possessed of unillusioned self-sufficiency. One man, less a Don Quixote than a soldier-saint, picked up the fight. Today, he is a lone hetman with the largest beat in the world to cover. The important aspects of John Feehan's biography are contained in the questions and answers that follow, all of which have to do with dung beetles and their economic value for the cattleman. The literature on the subject is vast, but the actual experience in the field is nil.

John Feehan is a one-man Chamber of Commerce for the beetle in Australia. The lessons he gifts the U.S. should be weighed out on a jeweler's scale, they are that valuable. His answers to our questions make suggestions that Americans cannot ignore with impunity.

The CSIRO dung beetle project has still another milestone to its credit, those dung beetle colonies established between 1968 and 1982. The entomology people imported more than 45 species, chiefly from South Africa. Received in quarantine, all were released in climatically appropriate locations. Equally as important, records were kept and mapped and published together with a plastic overlay that placed the nearest city, village or station in proximity to the name of the species. For the first time in history this exercise established a unique data base that supplied an audit trail. More important, it constructed a computerized protocol suitable for almost any possible location on planet Earth. A directory for, say, *Onitis alexis* has its index in terms of illustrated figures and maps, length 20 mm, occurs in warmer climates, releases made between 1973 and 1984, etc. Details follow with staccato regularity, color, flight time, life span — the type of information an interrogator might demand from an errant suspect.

The Onitis genus seemed poised to hammer buffalo flies into submission, assisted by Euoniticellus and the rest of that genus. It takes a lot of type to list 45 species, a task that Australia's leaders attacked with zest. Of more interest to Americans are *Onthophagus taurus, Onthophagus binodis,* and on down the alphabet. The legacy was not enough to hold back the companies with huge PR budgets and lawmakers not literate in the subject.

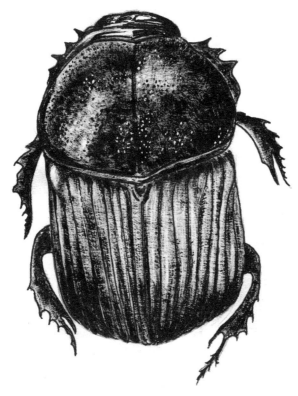

Onitis alexis *is a gift from South Africa via Australia in 1980. It has charcoal forequarters with bright sienna/chestnut wing covers.*

4

A Lone Hetman

The world has turned over several times since George Bornemissza proposed and guided the CSIRO Dung Beetle Project. Between 1968 and 1982, CSIRO entomologists sponsored the immigration of more species of the six-legged insects, not only from Africa, but also from Europe and Asian countries too numerous to detail. Once open to plant and animal importation without restriction, Australia by then had embraced scientific disease control with a vengeance. Those little animals were commanded into quarantine, then grown for two or more generations before being released onto grassland swards, always in clinically suited areas. John Feehan was both a progenitor and an inheritor of knowledge that awaits discovery in an Australian-American cattle population of at least the 40 breeds I have posted on my office wall — and others not accounted for — as well as an animal population — wild, feral, and domestic — still to be counted.

John Feehan chose to carry on after the Australian dung beetle program folded. Without the facilities to import and rear dung beetles, he adopted Homer's adage, "mind over circumstance." He set up a Website, opened a home shop for business, and proceeded to implement every lesson contained in the literature that had stacked up like cordwood over the previous 31 years.

"I know where the colonies are," he explains. He knows how to handle the species. He knows how to acclimatize them. And he has a monopoly on distribution now that the agencies of science have bowed out, except for occasional university probes. He doesn't look the part, but he is what I would call an intellectual accountant, a progenitor of the idea that we still have to develop an accurate accounting system, one that preserves the values. It would be "the greatest invention," if you will.

Listen to this totally pragmatic man. "One little simple exercise that I do is to take school kids into the paddocks by the busload. I get them to collect beetles. Believe me, 25 or 30 kids can clear a sizeable paddock of dung beetles in one hour. They're much better at it than adults.

"One little demonstration I do for the kids is to put six liters — a gallon and a half — of water in each of two buckets. I then locate a tunnel system in the grass and invite one of the students to pour the water on that spot. The gallon and a half disappears, like being sucked up by a sponge. Then I step away from the tunnel system and empty the second bucket. The kids all watch that water run like a stream toward the creek 20 feet or so away."

As simple as that demonstration is, it shows that the tunnel system perforates the soil. All those excess nutrients and urines that might be washed into a stream flow down into chambers and holes that have to be counted in terms of millions or not counted at all. I know of no scientifically validated study making this point so simply and dramatically. It takes resources, lots of tabular presentations, and a well-typed manuscript to make the point. Feehan chooses to make his points with kids who quite often make the same point with their parents. The premise is common sense ratified just the same. Water runs downhill. "You don't need a science degree to put that on the debit side of the ledger," is a summary remark by John Feehan. Nor does it take a science degree to know that something good is happening.

River catchment people are the object of Feehan's special antipathy. They want beef producers to double-fence each side of any stream that runs through a property. The select solution seems to be a grant of $40,000, perhaps $50,000, to do the fencing. Yet when dung beetles are on the job, melted dung, stray nutrients, excess phosphates, and surplus nitrogen simply can't make it to the water. A fence may slow the process, but soil without beetles will allow for water pollution anyway.

Water conservation and reduction of the parasite load became the goals as Feehan started his quest. He quotes at the drop of a hat the intelligence facts and figures to illustrate how beetles bury dung, taking parasites underground, and so entombing the mischief-makers that they can no longer make it to the surface. Complete rejection of parasiticides is therefore a sound syllogistic conclusion. "That, my friend," says Feehan, "is why the Monsantos, the Mercks, the Pfizers are not very keen on having dung beetles doing their work efficiently."

After approximately three decades of working with dung beetles, Feehan knows how to colonize beetles. First, he harvests the crop from sites where they are well established. This takes a degree of skill. Many producers are reluctant to allow anyone on their farm for the purpose of taking away their dung beetles. Feehan disarms his objectors quite simply.

"About 80 years ago, this country was overrun with rabbits. Farmers used to say, 'For every rabbit you see, there are six you don't see!' Well, for every dung beetle you see, there are 100 underground in the pipeline." Feehan assures farmers that taking 10,000 beetles off the farm is not going to significantly affect their population. In return, Feehan bestows a colony of a species that the farmer doesn't have. The barter system works quite well. No taxes accompany such an exchange, for which reason farmers like the arrangement. It's like having a few extra invisible steers that government gumshoes can't count.

Feehan's hatchery is the inventory of locations where the beetles are, where a deal can be made. He has his cleaning process, and he computer-matches species to climatically appropriate areas mapped out in *Common Dung Beetles*. Beetles are then sent out to a new home as a starter colony of 1,500. "Everything has a threshold over which it will build up and multiply," Feehan reminds. "With dung beetles, I feel it is 1,000 if the females are full of eggs. To make sure they colonize, I add another 500."

The lone hetman now offers to all farmers and beef producers a beetle-identification service. Farmers who want to colonize know the drill. Beetles need their conditions, the chief one being food. The requirements are almost as exacting as the checklist for an airplane before take-off. Failures are guaranteed if pesticides, fungicides, drenches and poisons are still viewed as miracle cures. In the outback, potential sites for colonization are hard to come by, not so much because of toxicity but because the rain gods are away for the season.

A half million tons of cow dung are dropped each day on Australia's cattle spreads. When dung pats remain exposed to sun, rain, wind and hatching insects, the resultant fly population literally explodes. The buffalo fly is reputedly worse than the American horn fly and face fly put together. The female sucks blood at least 30 times a day. Entomologists figure the flies torment a bovine animal with armies 3,000 strong, even 4,000 or 5,000 flies per beast. These succubus insects deliver a level of torment that goes beyond instant comprehension. It takes only five or six days for the buffalo fly to mature. If, however, the dung is buried within 24-48 hours by dung beetles, that physical maneuver becomes an effective birth control mechanism.

"It is so simple that school children can understand, but for politicians, it's all beyond their comprehension," mused Feehan. He has no illusions about the technology proposed by the university, any university! Insecticides work for a while, but then insects build up resistance. The trade then moves on to a more powerful formula. In the final analysis, human beings are subjected to chemicals, usually absorbing them in their fatty tissues. The human load is now so heavy, we are prompted to approve the cynical remark, "In a cannibalistic society, we'd all be unfit for human consumption, using Food and Drug Administration guidelines."

The best estimate says that 44 of the dung beetles colonized by the CSIRO have survived. Feehan has found 31 of these species alive and well in their assigned locations, and propagation on farms is proceeding quite effectively. Beetles range far and wide unless the presence of toxicity interdicts their progress. The lone hetman figures that dung beetles would bless all of Australia except for the toxic overload. "But it might take 100-200 years."

The colonies that Feehan has exported have all gone to research institutes. Three of his species have gone to the United States, *Onthophagus gazella* included. Although U.S. authorities are cautious in the extreme, "They need not worry about dung beetles bringing in unwanted bacteria or viruses. We have nothing in Australia that you don't already have in the United States."

When beetles find nasty things in the dung, they promptly put it underground. Once tunnel residence is achieved for the dung, micro-bio-activity takes over. The nasties are anesthetized. When an outbreak of some sort looms, dung beetles are Nature's living solution. They bury the dead and degenerated. They are certain to produce more benefits than negatives.

These things stated, it still seems obvious that a nation-continent the size of the United States is a bit much for a one-man dung beetle colonization distributor. John Feehan admits it. "I can't handle the job alone," and with that, Feehan started counting the 20 colonies being shipped to farmers around the country. Children and part-timers are all that remain of a dung beetle program that was once the envy of the world. A dung beetle program based on government money brings many people up fighting from behind their desks. In the U.S. the laconic apostrophe is, "They're paying farmers not to farm!" And the rejoinder is, "Really? Do you get a government pension?" The nod is in the affirmative. "Then you're being paid not to work!" The dung beetle customers served by John Feehan pay their own way in a world seemingly hostile to their endeavors.

John Feehan and Truman Fincher became as blood brothers during the zenith of dung beetle propagation under official auspices. Both had a passion for their grand objective. Both were destined to meet a similar fate. Both men have been required to observe that a similar level of inefficiency exists in both countries. Much the same is true of soil problems, ergo domiciles for Nature's clean-up crew.

Vast areas of Australia are well below sea level. Winds over millions of years have been blowing the continent into the sea. Any realistic attempt at accounting would sound a bell, but as a great stock broker once said, "There is no bell to tell you when the show's over."

Given enough resources, it would be possible to build the soils of Australia. Select farmers in the nation have proved as much. Dung beetles are the great enablers. When they were imported from foreign climes, especially from Africa, Turkey, and Asia, they had already proved their worth for millions of years. That is why farmers of 40 centuries were so sustainable.

A researcher in Africa put a marker on the ground, another the next year, and so on for five years. It was a radio-active isotope for correct identification. At the end of five years he was able to find those markers down to a depth of one foot. This was Nature turning topsoil all by itself. Indeed, Nature has been doing just that in Africa for millions of years, and now seems to tell us, "Go, and do likewise."

During his first dung beetle career with CSIRO, the modus operandi was simple. If the objective was to colonize dung beetles in the outback, or in Brisbane, or in any other area, then teams would find a like territory, usually in Africa, but also elsewhere, that imitated their target climate and terrain. It was an insurance policy for survival. For this reason and for reasons stated in the previous chapter, dung beetles are still firmly established from the tropics in the north to Tasmania hugging the Antarctic Circle. Cold climates have their own species, as do dry, wet and hot areas. The sort is in the details. One of Australia's colonized species can be found around Washington, D.C. and as far north as New York. A foot of snow does not disturb its welfare.

The world does not live at the level of its impassioned men. Others join sooner or later, hopefully before fatigue takes over. John Feehan tells me that he has cancelled his Website. He's too busy filling orders and preserving the fire of a program that ought not ever to go out. Fan mail is often no more than fan mail, but the dung beetle farmer is a national treasure that the world will appreciate, if not too late.

To paraphrase Yogi Berra, "The show ain't over till it's over." A codicil to John Feehan's story is contained in a letter that came my way as a copy. A part of it is quoted. . . "Tasmania has only 4 species which could be harvested, whereas the mainland [of Australia] has 30 introduced species firmly established on cattle stations in Queensland."

Feehan goes on to relate how he had sent two colonies to the U.S. and assisted with two others. Other species were being sent to institutes for research work. Feehan noted that the U.S. has the same problem with the education of the urban population. "I recently showed a live beetle to a TV presenter. She screamed when she saw it move. The rural-urban relation is exactly the same in Australia. Since leaving the CSIRO in 1993, I have given 150 dung beetle presentations or seminars to cattle producer groups, schools, land management groups, urban groups, etc. all over the country. Beetles have even been released in major cities to bury dog dung. I would like to think that 60% of the Australian population now views dung beetles in a positive manner."

Using a flotation technique, Feehan recovers dung beetles, then transfers them to field colonization sites. He identifies dung beetles free of charge for ranchers. He testifies that he often finds a single dung deposit alive with a thousand beetles. When an area is correctly populated, the dropping can disappear within 48 hours. Feehan's talk is entitled "Dung Beetles: Past, Present, and Future." Sadly, he relates that in his lifetime as a dung beetle activist and presenter, he's had only 1, 2, or 3 politicians who even listened. Still, he is as upbeat as any athlete on a comeback trail. If the government won't help, then people will have to help themselves.

"Help themselves" is exactly what is going on. The Australian Museum runs a program called Dung Beetle Mania. It encourages communities and schools to learn all they can about Nature's disposal crew. Learning about which species are in the home area is front-burner stuff. On a private citizen basis, a distribution system is indeed afloat with John Feehan as guru, operator, doer and decider. He may be a lone hetman, but to those who have dung beetle colonies established on their land, he is a pioneer, possibly a prophet, perhaps a Messiah.

It has been said that East meets West in Australia, and the United States is somewhere in between. That intellectual flow returned to the United States as if to answer the call of the beetle.

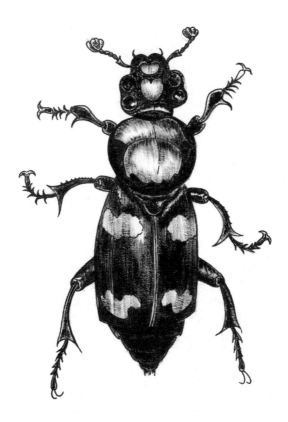

Necrophagous americanus, *a famous American burying beetle, and often a resident of frontier graveyards, is black with a slight granulated texture. Its patches are light orange, almost "schoolbus yellow."*

5
The Call of the Beetle

No one could visualize the Australian dung beetle world ending not with a T.S. Eliot bang, but a whimper, not when America's grassland farming was coming wide awake. Even at the time of Henry Jackson Turner, it was clear that farming in exploitive hands was coming to an end. Officially the frontier was closed before the arrival of William McKinley and Theodore Roosevelt in the Oval Office. By then, land was no longer available just over the horizon. The idea of break it out, use it up, and move on was rushing the country toward a dustbowl, a great depression, and an age of chemistry that has still to run its course.

By the time the country discerned that it had only to "fear itself," a movement was on the way to rediscover practices that some farmers had been using successfully for 40 centuries. Time-tested practices such as crop rotations, sheet manuring, conservative grazing and practical breeding finally came to academia and other intellectual advisors like a delayed flash of insight.

If new ground was no longer available, then old ground had to be refurbished and productivity restored. During the 1930s, the term "grass famer" became current coin. Edward Faulkner's *Plowman's Folly* and Louis Bromfield's books reached a high level of recognition. Though the word "sustainable" had to wait a few decades more to achieve respectable usage, sustainable rather than extractive was on every political tongue. Articles by Missouri's William A. Albrecht were published in dozens, if not hundreds, of agricultural journals of the day.

By the late 1940s, real progress was being made informing farmers as to the possibility of soil conservation. County agricultural agents and the farm press stressed the value of soil organic matter and ways to increase it. Proper use of manure was recognized as being of both economical and ecological value. A well-run farm was seen as having a number of crops planted and harvested at different times of the year in order to spread the workload. Pastures were used to build soil productivity and to feed livestock which were used to utilize crop residue.

On the best of farms, there was no such thing as waste products. Inputs from outside the farm were minimal. This era featured that elusive phenomenon, the family farm, but during the last half of century 20, agriculture has changed from a biological operation to an industrial procedure. It was the explosive growth of the chemical industry caused by World War II that changed the very nature of farming.

With cheap and abundant nitrogen, the role of manures became downgraded. Problems surfaced almost instantly. Academia figured that the problems of technology could best be solved by more technology. In less than a generation, science replaced centuries of accumulated knowledge. The results have been tragic. Soil came to be regarded as an anchor for plants, nothing more, not as a complex association of minerals, organic matter and innumerable organisms, all vital to our survival. The losses

have been concealed by faulty accounting, by the development of genetically modified crops and by chemicals of organic synthesis. Admittedly,
crop yields increased under the new regime and this made Nature's own
accounting process even harder to recognize.

Fishing through my accumulation of notes, I found the following in
a yellowed reporter's notepad. I believe it was put on a tape for me by
William A. Albrecht before he passed away in the early 1970s. "We have
been poorly served by the scientific community which has proved the
short-term benefit of each new chemical and each bigger machine, all
without regard for the long-term effect. The day we find that our soils are
dead and unable to perform their function as water reservoirs so that both
floods and droughts are more common and severe, we will have lost the
biological diversity that once held pest organisms in check. Insects, pests,
weeds and diseases of both plants and animals will become even greater
problems. The use of pesticides has increased manyfold, while crop losses
have more than doubled. Concentration of livestock in factories has created waste disposal nightmares while robbing the soil of nutrients and the
organic matter of manures."

Microorganisms, bacteria, fungi and mycorrhizae turn manure into
rich, healthy soil, soil that can push up nutritious grass. Dung beetles are
really Nature's own "sanitary engineers," quoting Pat Richardson. We have
become used to believing that species evolution always requires thousands
of years, but as early as 1962 it became a matter of record that 10 of 33
beetle species in Lancaster and Cheshire, England, starting at the beginning of the 20th century and tracking them up to the 1960s, had disappeared. Only one species answered the call of discovery. Some species
were discovered in fossil samples, which suggests stability for thousands
of years in spite of climate change and ecological traumas. The core species remained the same for 43,000 years, yet there was discovered in 1962
a sudden decline in species in hardly 50 years.

It appears that different rules apply to common and rare species, a
paradox. Did changes in cattle management proscribe non-core species?
Decline can be measured. Work in the early 1930s, since enlarged, tells of
the decline of the dominant species from 83% to 30% to 14%. The dominant species do not always remain dominant. The recent domination of
Onthophagus gazella, after a period of interdiction, is a case in point for
both the U.S. and Australia. Vegetative changes often add to or subtract
from the population of a group.

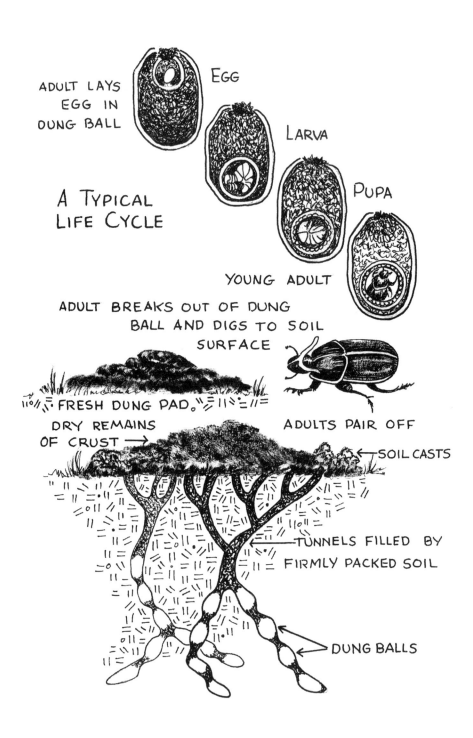

ADULT LAYS EGG IN DUNG BALL

EGG

LARVA

A TYPICAL LIFE CYCLE

PUPA

YOUNG ADULT

ADULT BREAKS OUT OF DUNG BALL AND DIGS TO SOIL SURFACE

FRESH DUNG PAD

DRY REMAINS OF CRUST →

ADULTS PAIR OFF

←SOIL CASTS

←TUNNELS FILLED BY FIRMLY PACKED SOIL

←DUNG BALLS

"What we have here," the sadistic warden told Cool Hand Luke, "is a failure to communicate." Well, what we have here is a failure to immigrate, sequestered colonies presiding over common species, change and stability in northern colonies over thousands of years, yet there are shifts due to range practices. A rare species becomes extinct, much like an individual Einstein, first migrating, then dying off as procreation is terminated — in short, satellite species court extinction.

Many species of Scarabaeidae dig vertical and lateral tunnels under a dung deposit, and then transport their treasure to the bottom of a burrow. Their harvest can serve as a breeding chamber or a gourmet meal. Life from the brood ball depicts Osiris, Horus, the Sun, the god of the dead, and the panoply of Egyptian mythology as symbols more enduring than brass. Symbolically, Osiris and Horus are in the brood ball. The dweller tribe prefers large droppings, possibly because they like to swarm over a new deposit like Vandals entering Rome. As a lower species, the dwellers have great fecundity, 100 eggs or more being typical.

The first stage of any beetle is to a choreographed purpose, to eat! One and all, they eat their way through several stages. Dung beetle development is rapid, quite a different scene from their distant relatives, the cicadas, which take years to mature, then emerge from the soil, hang out in trees and deliver a symphony that may or may not deliver I.Q.-inhibitor sound waves in special areas such as Washington, D.C.

Nature's process is so special that it shames the accomplishments of those who presume to genetically engineer life. It is probably correct to say that the pupa is the medium for the bio-chemical transformation of this eating machine into a larva, then a breeding unit, the adult, which is governed by the foibles of its kin. Egg, larva, pupa, adult, all of Nature's beetles travel this road, our ignorance trailing along without the foggiest notion of why.

Nature, we say, is fecund. Yet there are dung beetles that lay a single egg, whereas others seek to explode the population, as illustrated by a cluster of brood balls on page 48. The female does not kill her daughters, as is the case with the queen bee, which apparently seeks to remain queen of the hive. Even the dung beetle single-egg producer is courted by a bevy of males.

Some species leave their eggs in the soil or deposit them in static compost piles or dung heaps. If we depart from dung beetles to look at the whole pantheon of beetledom, we open a diversity of Nature larger than those discussed in Charles Darwin's *Origin of Species,* wood eaters, forest dwellers, carrion seekers, flower pests, even the cotton bollworm and luminescent firefly qualify, but all are outside the purview of this volume.

These two nests are the work of two females. The one on the right is complete, the other still in a construction phase. The ventilation plug is made from fibers in the dung. There is a partition between nests.

No data base has ever been constructed that contains all the organisms that science has uncovered. After all, it has only been two centuries since Linnaeus made his mark. Since then only a few smaller groups have neared completion. It is said that the inventory of birds is quite stable. There are about 9,000 species. The count for mammals is now at or near 4,000. Of all life forms designated as animals, insects seem uncountable. "We simply do not know," E.O. Wilson noted in his seminal work on biodiversity. They hide out in caves and many other places, these insects called beetles or weevils. They live and die underground. They often defy classification and have entomologists at odds and sometimes at each other's throats over the classification method. Not a few remain in limbo, blind, flightless, as if awaiting evolution's signal to go ahead, separate from the present life to engage in the time and eternity that evolution seems to demand for change of status and species. Weevils, by the way, of the family Curculionidae, are only one of about 150 families of beetles worldwide.

Darwin notwithstanding, the fact remains that beetles exhibit very little evolution during the 500,000 years that common agreement says man has been present, having arrived by a route forever to be debated. Fossils recovered from that era are identical to present fauna. Many of those prehistoric specimens have been pulled from the La Brea tarpits now surrounded by Los Angeles. Two species presumably extinct, *Onthophagus coerestae* and *Copris pristinus,* have been paired with modern species in Texas and Mexico. Their entombment together with prehistoric camels, sloths, mammoths and herbivores suggests a working connection with bison herds that once populated areas as far apart as mountain meadows, plains, even woodland pastures of Pennsylvania and parts of the South.

Even a decade ago, beetle trackers listed 378 species as globally endangered, vulnerable or rare. Even rarer is adequate information on this grand diversity. Ants have their biographer in Edward O. Wilson of Harvard, but dung beetles seem to have not been even taken for granted. The world may little note the disappearance of beetle species, but it can hardly escape the fact that another important species is also disappearing, the coleopterist.

The little package that frightens ladies and enchants children is simply an exoskeleton as skinless as a turtle's top, its three basic parts are as defined as the fingers of a baseball glove. The many parts sound like the innards of a Chevy truck, but the body plan of the skinless, shelled creature that has summoned our attention is not very different from the armadillo, tortoise, terrapin or swamp turtle. Indeed, one variety is called the tortoise beetle. The dung beetle is possessed of the usual eye-bearing head replete with antennae and a marvel of a mouth — excuse the exceptions. The basics are stated in the line drawing in Chapter 3.

Dung beetles have been compared to earth-moving equipment on a new construction job. Look at them under a strong reading glass and you'll see a high loader, even a modified drag-line, tamping unit and bulldozer. This isn't hyperbole. Actually, it is safe to say that dung beetle engineering magnified has made subways and channel construction possible, once the state of the arts picked up on using Nature's slide rule.

Kara is the name of a granddaughter who came by as I puzzled over what to say and what to leave out in telling farmers about this grand and glorious bug. She had been to an eye doctor.

"Do dung beetles have eyes?"

"They certainly do, sometimes more than our alloted pair," I said.

"Do they get blepharitis?" She exhibited the eye treated by a physician that morning.

I didn't know then and I don't know now. I could, however, relate the intelligence that some beetles were blind, evolution perhaps closing down eyesight as being of no consequence to those forever in the dark. But this is not the case with those flying beetles that trip from cow pie to cow pie. If they have eye problems, they're on their own, never the patient of a beetle physician. I told the little granddaughter about the three segments of a beetle, the head, thorax and abdomen, all held together by a chitinous membrane. But you can't divert a child's attention with a politician's departure from the point in question. "Some beetles, not necessarily dung beetles, have quite a few eyes in a row."

"I hope they don't get blepharitis in all of them at once," she said, and then she was gone. She didn't hear about the shovel, scraper and high loader features of some dung beetles, or about the abbreviated list of features of this wonderful little animal. Cave dwelling beetles often are blind or have minimal eyesight. Subterranean dwellers live off bat guano and such other pellets that come their way, always protecting their delicate bodies from desiccation. Some types living in Montana caves appear to be distant relatives of similar species in Europe.

The American burying beetle is not a scarab, even though it shares with some scarabs a penchant for carrion. Its family name is Silphidae (remember that all family names end in —ae), and falls under a superfamily name, Staphylinoidea. I make this distinction because the illustration at the head of this chapter departs a bit from Scarabaeidae, the focus of the inquiry. This American beetle is a shirt-tail relative of the family Agyrtidae, a beetle that shares with the condor a love for the dead.

The burying beetle may be the most studied of all beetles because of its comfortable existence in the laboratory. The literature is chock-full of papers that tell how to sex beetles, the conditions for successful copulation, brood establishment and the art of sequestering adults. As a matter of species preservation, university laboratories have been equipped to propagate and serve up for release populations in the natural environment. Suffice it to say, the American burying beetle is the giant among carrion handlers in the United States, sometimes up to 35 millimeters in length. There was a time when it was known as the sexton beetle. It often hung around church yards (cemeteries) the way a sexton hangs around the bell tower. Its role may have been diminished now that the human burying ritual requires a hermetically sealed coffin, but there are always bird and animal carcasses that require interment. This recycling chore proceeds quite unseen, which may or may not explain why sightings of this credentialed scholar of the graveyard are so rare.

Removal of dead animals is a must if some semblance of insect control is to be achieved, but few collectors find this little animal. As with dung beetles that also accept dead animals as gourmet fare, the American burying beetle prefers woodlots and deciduous forests as a habitat. Less than three decades ago, it was the collective wisdom of coleoptorists that americanus had localized in Rhode Island, Oklahoma and Arkansas. A spotting in Nebraska suggests that that undisturbed habitat may still harbor mature populations of this natural undertaker. Another beetle of the burying ilk is called *Necrophilus germanicus*. Indeed, the craftmanship of such beetles is exhibited in other nations under other names. In Japan, it appears that appreciation of this style of beetle accounted for preservation of fast disappearing forests. The American burying beetle became an entry on the endangered species list in 1989.

Coleopterists, of course, have examined the anatomy of all beetles that have been found and identified, always astonished at the grand diversity of Nature's designs. One beetle that could comply with P.T. Barnum's criteria for showability is the featherwing beetle, *Namosella fungi*. It is found in the eastern United States, and can't consume much of anything with its undersized eating apparatus, a 0.35 millimeter body and other machinery to match. It is so small it can navigate even the eye of a needle. The details intrigue the scientist and cause the eyes of cowboys and children to glaze over when they are recited in detail. Still, the social contract that the several species have, and their arts for survival from their enemies are as fascinating as, say, the study of ancient Greek culture or the intelligence that Democrates postulated the atomic scale in principle, if not in the detail furnished by Mendeleyev.

Much like many of life's counterparts, beetles have been endowed with four life stages, egg, pupa, larva, adult. Eggs are as soft and pliable as well-cooled Jello. Water penetrates the shell, as does a measure of oxygen — in some cases! How can one generalize when species number in the thousands? The word "some" may sound like weaselish hedging, but it has to be used, for there are beetles so small they can crawl through the gap in a tightly woven reed basket and others as big as a rodent. The jaws, or mandibles, of our most favored dung movers are so powerful that — raised to the size of a pit-bull dog — they could outpower even that terror of the streets.

Horns, antennae, peacock colors and variations could fill an encyclopedia. The horns on beetles sometimes have dilettantes, even professionals, jumping to conclusions. The horns are not fighting weapons, in fact, they're not even very functional. Perhaps they represent some sort of evolutionary failure, a left-over appendage chiefly of value to collectors, repositories and museum drawers. Perhaps horns are obscure because we don't know how to read them. Not many cowmen know how to read the horns on a herd sire, so continuing ignorance about beetle horns — origin, reason for being, possible evolutionary goals — remains one of Nature's mysteries.

They are nocturnal, these horned beetles. If they do battle in some way, they do it in the dark. Those small racks may play a part in defending food or in mating games, perhaps in the poker front exhibited to friends and enemies alike. *Onthophagus taurus,* mentioned elsewhere in this text, even takes its name from the victim of the bull-ring. And what shall we make of pushing contests that occur between male and female?

No one can say that dung beetles are good flyers. When their encased wings are uncovered like some secret weapon in a military silo, they rise up almost helicopter style, then lumber along like an early Wright Brothers plane. Flights are usually short, always furtive, more like an ancient cargo plane, not a sleek jet-fighter.

These little animals exhibit so much diversity, they almost challenge the imagination of all science fiction writers combined. Some harvest toxic materials from plants and store them in their own body structure, courtesy of leaves and plant saps. They then use these materials for self-defense. There is an operative chemistry in beetles that collect plant toxicity that would make airport terrorist-hunting fools wretched with envy when dreaming up new dangers. The list of chemicals that these little animals find or compound might fill an apothecary manual if collected. Even a species of longhorn beetle gathers the essence of the nightshade plant to be delivered as a protective fluid missile when needed. Blister agents and pesticides are sometimes produced, just as chemicals and compounds up and down the scale emerge on the worksheets of the chemist in his or her analysis.

They are chameleons, these connoisseurs of night soil, human and animal. They can defend themselves from predators by feigning death or by wearing camouflage colors. They can hold fast so tenaciously that collectors can barely dislodge them, or they can assume the appearance of an inedible substance or of an enemy known to Nature's world as toxic and lethal.

Species only distantly related to dung seekers have either caused or endured evolution that turned their bodies into a chalice of poison, a mini atomic bomb that keeps all predators at bay. Some are better mimics and impostors than Lon Chaney. It is probably safe to say that the various families have members that baffle the fools and fool the wise, humankind included. As with the law, investigators often read more into beetle behavior than they read out of it. Five-star dining, after all, has to do with palatability, and beetles unpalatable are simply beetle non grata to predators.

In our black and white print world we tend to leave unstated the fact that many beetles, even dung beetles, are so colorful that they enchant amateur and professional coleopterists alike. Are the most colorful merely advertising their prowess and exhibiting the statement that they are captains and kings of their world? Or are the colors bravado, a sort of "bring-'em-on" gesture? The mimickers take note and try to pass themselves off as the real thing. With man's jaundiced eye, we wonder aloud and in silence, fully aware of the fact that our science merely scratches the surface even though imagination runs wild.

In the high, off-desert area of the Santa Rita mountains southeast of Tucson, Arizona live no less than three species of scarabs, including *Plusiotis gloriosa* and *Plusiotis bejera*. Coleopterists describe them as beautiful, and collectors would glory the day when any one of them could be added to their collections. Not one of these animals could live in the desert a few miles below those mountain ridges, escarpments and mesas. The point here is that a mountain peak in Arizona, California or Washington is as isolated as an island in the Pacific. One can imagine the challenge to beetles of any stripe living above the timberline. Dung beetles have to be the specialists suggested in the chapter entitled Five-Star Dining. We do not think of dung beetles surviving under snow and foraging for food atop a sheet of ice in the warm part of the day, but this they do. How can there be an absolute when an exception is encountered at every turn?

So there is really little to shock the delicate among us when beetles slurp dung, house their eggs in dung and gift valuable dung to the soil's micro-organisms. The dung beetles duplicate man's art of composting by carrying the fecal material to the micro-organisms, rather than inserting microbes into the compost pile. Long before it came to *Homo sapiens* that to conserve the resources of the earth they ought to recycle, dung beetles were doing just that.

Quite a few dung beetles are specialists. The literature identifies them as no single venture into the wilderness could, or even one coleopterist in one lifetime. Tree dwellers in tropical forests often descend to earth to defecate and bury their dung, apparently preferring to gift soil microbes with their treasured excreta rather than allowing ever-hungry dung beetles such largess. Or do they really cover up their leavings to prevent desiccation, this with the assurance that dung beetles will find the cache? There are answers, and answers are sometimes contested.

In Australia, coleopterists have identified no less than six species of Onthophagus partial to wallaby and kangaroo pellets, fare that America's adopted *Onthophagus gazella* would reject the way a dog rejects a plastic biscuit. These beetles have served down-under wallaby dung so expertly that its accumulation rarely remained on the surface long enough to draw flies or invite parasite incubation.

Scientists talk about all beetle species, which now number somewhere between 290,000 and 350,000 described during the last 150 years, and so dung beetles are often dismissed with et cetera. Linnaeus started the ball rolling by describing 650 species. Two hundred fifty years later the classifiers are still in a quandary. The co-discoverer of an ancient Greek idea ultimately named evolution, Alfred Russell Wallace, early on proposed classifying beetles — dung movers and otherwise — according to habitat, tropical or temperate. It was and remains a capital idea. The native dung beetles of temperate zone America, even some out of our subtropical region, perform well in dealing with rabbit and deer droppings, but fail a passing mark when confronted by excreta from herds of cattle. How Temperate Zone dung beetles coped with the droppings of vast bison herds is still being studied together with dung beetle species that may have become extinct without leaving a trace. A lot can happen in 230 million years, making beetle genealogy a daunting task.

It was into this world that George Truman Fincher stepped when he entered the University of Georgia, ultimately receiving his Ph.D. there in 1968, and became a preeminent scientist, scholar and promoter of dung beetles.

Gromphus lacordairei *was imported from Argentina in 1985 as part of the College Station, Texas propagation and colonization project. It is a basic medium brown with a slight rust cast, smooth and shiny.*

6

Get to Know Truman Fincher

College research, funding and personalities make the business of reading Nature quite human. This was well illustrated by James Watson in that injudicious book, *The Double Helix.* As the world knows, Rosalind Franklin was shared out while her hostile associate, Maurice Wilkins, routinely made the run from Kings College to Cambridge to assist Watson and Francis Crick to Nobel laureate status.

George Truman Fincher came to academia via the usual post-World War II route, farm to school, University of Georgia, major entomology, minor ecology. The choice squared well with his life-long interest in bugs and nature. He was an excellent student given to puncturing the rainmaker's expectation. He accepted the kudos and applause that came his way as "too much pepper."

In those post-war and post-Depression days, old-timers recalled the preacher who came to town to pray that the Lord soften the hearts of the more affluent brethren. He prayed that they send barrels of flour, barrels of sugar, barrels of beans, barrels of pepper…"

"Hold on!" shouted a wizened farmer, "That's too damn much pepper!"

Truman Fincher never liked "too much pepper." He planned his work and life and worked his plan. He took his Master's Degree in 1966, the disciplines being entomology and microbiology. Some of his undergraduate work had been in entomology and ag economics. There was foundation work between 1957 and 1959 at Abraham Baldwin Agriculture College. Long after his name started to appear in various *Who's Who* volumes, and awards of merit became his personal property, he saw satisfaction only in his passion and dedication to the quest that has come to be synonymous with his name.

After receiving the Mallinckrodt Lifetime Achievement Award, he moved on to become a presenter at meetings of professional societies as well as before farm audiences. Ranchers and farmers saw him as a metaphor for all the workers in the dung beetle vineyard, starting with investigators as far back as the beginning of the 20th century. He was recognized as such when CSIRO invited him to Canberra, Australia to exchange information and facilitate the progress of two grant programs. Similar exchanges were taken to Germany, Canada and Argentina. When confinement feeding became college-recognized state of the arts, Fincher was on hand to spell out the consequences, all experimental expressions of pasture failings not relieved by dung beetles.

Truman Fincher had already found out that a significant grant had been approved for him. This grant was transferred willy-nilly to College Station, Texas. "They took my grant, and I raised so much hell that it was canceled," Fincher explains. He found out later that money was not available for work on insects to anyone not engaged in entomology research.

At the time, Fincher was involved in parasite research, and for reasons not clearly explained, parasites were not classified as entomological fare.

Much the same was true when the debate got down to considering a management headquarters for "dung beetle initiative U.S.A.," or at least the southwest part thereof. Some of the scientists had first-hand knowledge of the facilities at both Tifton and College Station. Might it not be a good idea, suggested Truman Fincher, to conduct a well-coordinated program for review of both locations?

So far, Tifton had the most research, the best tested theories and a natural laboratory of off-shore islands with livestock and native beetles that seemed to argue against the worthlessness of natives. Could such observations be coming from Texas? The foremost taxonomist, R.E. Woodruff of Gainesville, Florida, was in Fincher's corner. The foremost grass breeder, Glenn W. Burton, another Tifton collaborator, made the point that the Tifton station was also strong on agronomy, soils and water research.

Tifton, Georgia lost out as the site for the introduction lab because scotch got in the way. Texas had a loftier criterion. With the decision made, Fincher moved to College Station in 1978. The scientific application of dung beetle research to the cowman's profit profile really got underway after that. The task at hand was no armchair affair. Fincher spent a great deal of time in Argentina, always returning with caches of eggs for propagation release and colonization. They were large dung beetles capable of handling large quantities of dung. They were released with *Onthophagus gazella* and other African species near Hugo, Oklahoma. A rancher named Roy Catulla had 2,500 acres with dung to spare. He and other ranchers, along with Extension and County Agents, were all cooperative. Georgia and Texas also had release sites. All the beetles were colonized under quarantine in the lab, then colonized again in the field. Six of the species, led by *Onthophagus gazella,* proved quite successful, as Walt Davis and his neighbors soon found out.

The areas cut out for colonization by Truman Fincher and his colleagues had a continuity to them. Dung beetles do migrate, for which reason each site of colonization became enlarged, the sole governors being the food supply and the cowman's penchant for putting a bovine animal on every usable acre of ground.

Fincher and fellow lab workers also worked with California scientists, shipping out-of-quarantine species to them. The traffic to and from Australia was brisk. *Onthophagus gazella* proved to be the fastest responding dung beetle, the long-odds pilot for any release, out-performing the other five or so species in most releases. Live beetles and brood balls were proposed to the pastures, leaving it to Nature to do the disposing.

The tunnelers proved most valuable. In fact, rollers couldn't function very well in lush pastures, and, like the female flea, moved backwards to complete their task only with difficulty. Beetles capable of rolling a dung ball after a camel passes in the desert would prove marginal if not useless in an ecologically sound pasture with enough grass to feed a cow her 3% of body weight each day. Retreat via flight more often than not identifies a frustrated roller.

"You'll be called to account for every good deed you've ever done," the bad man in the typical cowboy movie might say, and so it was with Truman Fincher. He was told to quit releasing dung beetles. His research was redirected to the study of predators, this when the real nemesis of the horn fly was the dung beetle.

There was a lab review in 1990, Fincher now recalls. The number two man, an associate administrator, chaired the affair, the main objective being to sack Fincher's research leader, Robert Harris. It was generally agreed that Harris should have been removed years ago. Lab scuttlebutt had it that his appointment to that position had been made because Harris was once a neighbor of the man who made the decision.

"The national program manager was a 'tick-man' named Ralph Bram. His expertise ended right there," Fincher explains. Bram was an intractable foe of Harris. Bram transferred him to the screwworm program, one of the most successful ever launched by the Department of Agriculture. The secondary objective was to exile Harris to Mexico. He'd refused to take early retirement. When the associate administrator of the Agriculture Research Service came down for a program review at the lab, a question arose about the dung beetle program.

"I thought we got rid of the dung beetle program," came like a shock wave from Dan Lassiter, the gavel holder. He'd meant the research leader problem, not the work horses. Bram misunderstood. When he returned to Beltsville, Maryland, he called Fincher's new research leader and told him to stop all dung beetle research. Concerning the 10-12 species then in a quarantine colony his instructions were "to kill them."

Fincher demurred. He found two foundations in Oklahoma willing to take the beetles and salvage something from the expenses tied up in dung beetle research and propagation, but neither had the expertise necessary to maintain and further propagate the insects. One wanted to install the trays in a building that had previously housed chemicals. Such lodging was an insurance policy for termination of all beetles in a few days. The Kerr Foundation worker left six months later, and, with her departure, the beetles simply died off. Fincher carried on privately. He continued to release them even while working with predators. The speaker had said "problem," and the Beltsville bureaucrat had thought he'd said "program."

The show is never over until it's over. The bureau people next got rid of both people and money. The usual procedure is "early out" retirement. A 50-year-old scientist with enough tenure could refuse a transfer. Fincher's associate, Richard Blume, refused a proffered transfer and retired. Fincher himself refused the offer of a transfer to Sidney, Montana.

Readers can be allowed to speculate about forces hiding darkly in the wings, as Thorstein Veblen put it, about the grab for dollars, about internecine strife, even about a research director's personal repugnance toward dung beetles. When the last of the budget money evaporated, so did the grand experiment, the only legacy being colonized dung beetles that survive to this day and a hope that farmers would step forward to continue the work much as John Feehan is doing in Australia.

As the work progressed for this biography, it came to a thunderclap conclusion with the following obituary.

The USDA-ARS Biological Control of Horn Fly Project began in Kerrville, Texas in 1970 when Dick Blume began research on dung beetles as biological control agents for the dung-breeding horn fly. Dung beetles compete with horn fly larvae for the same food resource (cattle dung). In 1970, Blume received 2 exotic species of dung beetles from Australia (CSIRO Dung Beetle Project). He released the Afro-Asian beetle *Onthophagus gazella* in Texas in 1972 after transferring to College Station, Texas. During the next few years, with help from state veterinary entomologists, additional releases of *Onthophagus gazella* were made in Alabama, Arkansas, Georgia, Mississippi and Oklahoma. University of California scientists also released this species in their state, and it is now established in the southern tier of states from South Carolina to California and it was recently captured in southern Kansas. It has also dispersed southward through Mexico and has been reported as far south as Guatemala.

The USDA-ARS Biocontrol of Horn Fly Unit was established in College Station, Texas in 1977, and additional biocontrol research was initiated by three scientists on parasites and predators of horn flies. Fincher transferred to Texas in 1978 and began foreign explorations to secure additional species of dung beetles for our quarantine facility in College Station.

A South African dung beetle, *Euoniticellus intermedius,* was released in Texas in 1979 and, with help from UC scientists, it is now established from southern California to near College Station, Texas and southward into Mexico. A dung beetle from Europe, *Onthophagus taurus,* became established in the southeastern United States in the early 1970s, and thousands of specimens were released in Texas. This beetle is now established from east Texas to Florida and up the east coast to New York. It has also been reported in Tennessee, West Virginia, Pennsylvania and Ohio. UC scientists also established this species in California, as well as an African species, *Onitis alexis,* that failed to establish itself in Texas and Georgia. Another South African dung beetle, *Liatongus militaris,* was recently found on the King ranch near Kingsville, Texas, where it was released from 1985-87.

Fifteen different species of foreign dung beetles were imported and released in Texas and other states from 1972-1987. In 1987, as related above, an administrative decision was made to terminate the importation, mass-rearing and release of dung beetles and emphasis was switched to introducing exotic species of predators of horn flies. Since then several exotic species of dung-inhabiting predators Staphylinidae and Histeridae, have been introduced and released in Texas. One staphylinid species, *Philonthus flavocinctus* from southeast Asia, was found one year after release, but establishment has not been confirmed.

Descendants of exotic species of dung beetles released by the USDA-ARS Biological Control of Horn Fly Project presently populate millions of acres of pasture, often resulting in 100+ beetles attracted to each deposited cow pat. In such numbers, there is total destruction of the dung deposit within 24-48 hours. These beetles have been shown to significantly accelerate cow pat degradation throughout much of the year with a resulting negative impact on the survival of immature stages of horn flies and gastrointestinal parasites of livestock. In addition, millions of tons of cattle dung containing valuable nutrients, e.g., nitrogen, organic matter are annually recycled effectively into the soil by the actions of these dung-burying insects.

On September 30, 1997, the USDA-ARS Biological Control of Horn Fly Project was terminated.

Truman Fincher was charged with preparation of the above-quoted encomium, with the provision that it be approved for presentation before a national assembly. At the last minute, Truman was told to go to the meeting, but he was forbidden to make the presentation. There had to be more to the story. Interviews and a few spare documents provided the rest.

Pat and Dick Richardson of Austin, Texas, along with the University of Texas have experimented with dung beetle farms in the image of ant farms. Holistic Resource Management clubs and meetings pass on the word, but a working mechanism for harvest and recolonization, under or above ground, has still to be developed.

Even before the New Orleans conclave came to the wrap-up stage, entomologists and those of allied professions had agreed that geological and anthropological events had deprived most of the United States of the dung beetle species capable of returning profit to cattle producers. Introduction of dung beetles harvested from countries with similar climate and soil types was more than indicated, it was mandated. The experts had a name for the mandated observation. It was the fact that successful coprophagous fauna had to include species that could live together in relative harmony. With thousands of dung beetles on planet Earth, the task became finding coprophagous types in the larval and adult stages. This meant pasture farms.

The checklist of questions, many of them without answers, became a part of the full report in New Orleans in September of 1976. It asked for the consequences that might attend the bold program. Indeed, would the proposals...

1. Reduce the incidence and severity of animal parasitisms, diseases and insect pests by removing feces from pasture surfaces?
2. Reduce the production costs of animals on pasture by lowering the need for fertilizer, increasing effective grazing areas and improving soils by the incorporation of organic matter?
3. Reduce the pollution of pasture surfaces by the burial of feces and allowing rapid decomposition by microorganisms?
4. Reduce or eliminate potential hazards from chemicals and chemical residues?
5. Contribute toward an improved nutritional level for mankind through increased quantity, higher quality and lower costs?
6. Reduce labor and processing costs (trimmings, condemnations) in slaughter houses?
7. Contribute knowledge and technology that would lead to a reduction in losses from parasitic diseases in all classes of livestock?
8. Remove hazards to human health by removing animal wastes which contain pathogens common to both animals and man?
9. Increase the esthetic value of the countryside?

Further, there were the stated research objectives:
1. Which dung beetle species would be suitable?
2. How many dung beetle species are needed?
3. What population density of dung beetles is needed?
4. What effect will an introduced species have on the native dung beetle fauna?
5. Will rapid burial of feces control livestock parasitisms?
6. What effects will burial have on the survival of parasite eggs and larvae?
7. Can dung beetles destroy parasite eggs and larvae by ingesting them with the feces?
8. Will the introduced dung beetle species be able to serve as intermediate hosts for native parasites or transmit bacterial and viral diseases?
9. Will the incidence of certain animal diseases be lowered by the introduction of dung beetles?
10. Will the same probable benefit of reduced parasitisms be extended to wildlife?

It may be correct to say that, on a world-wide scale, more students and professors since World War I had worked on dung beetle research than on any other entomological discipline. As early as 1922, far-seeing researchers imported dung beetle eggs from Mexico, a happenstance that made importation for soil improvement a politically correct topic for conversation.

By 1969, a one-way traffic between Australia and the U.S. had established itself, the chief objective being the importation of the African dung beetle named *Onthophagus gazella* and some few other species. By 1971 approval came down from on high to make experimental releases in Texas. None of this activity proceeded without first gathering available knowledge, especially from Australia, and from literature on the subject that was beginning to stack up like congressional speeches.

It was the popularly called "gazella," so named by splitting off the Onthophagus genus handle, that captivated dung beetle workers across the South. Releases in 1972-73, which opened the door to more releases, stepstoned across the counties, soon to be expanded into neighboring states, each release and its results duly noted and duly enshrined in the bibliographies of that current generation. The full story of that outbreak of ecological sanity caught the attention of some few popular press publications, for which reason the dung beetle advance took seven-league strides forward.

Imports tagged by field men arrived from Korea and Pakistan before 1975. Initial field releases in Georgia and in Colorado County, Texas were made that year, and in 1976 a quarantine facility at College Station, Texas was approved. The early 1970's saw the arrival of the African contributions, *Onthophagus gazella* and *Euoniticellus intermedius,* by way of Australia, usually in pairs of 100.

Now identification surveys got under way. The most advanced of the dung beetle pioneers must have foreseen a map similar to the one that Australia eventually published, each eco-climate area correlated with the dung beetles most likely to survive. Even the King Ranch in Texas got its fix of dung beetles in those halcyon days. *Onthophagus gazella* seemed to migrate best of all, often being recovered 20 miles away in Kleberg County, even 100 miles distant in some cases.

It was at the New Orleans meeting in September of 1976 that the judgment was made to set up a dung beetle insectary. It turned out to be a pivotal point in the career of Truman Fincher. He had completed his Ph.D. dissertation on dung beetles as related to swine parasites at Tifton in 1966 under the auspices of the University of Georgia. "They created a position for me when I graduated," Fincher later recalled. In 1969 the International Congress on Entomology met in Washington, D.C. Large scientific meetings are where friends are made and disasters kept at bay. Careers take strange new directions, not so much because of what the lectures contain, but because of side-bar conversations with colleagues. Fincher found himself talking with the lab director at Kerrville, Texas. When the intelligence that Fincher was working on dung beetles was passed on, it developed that Kerrville also was working on dung beetles. Truman went to Kerrville for an introduction to *Onthophagus gazella*, recently arrived from Australia. "We agreed to push for a dung beetle program," Fincher explains, "and I was trying to get it in Georgia."

Fincher's area director did not want it at Tifton. It would interfere with the golf course and scotch-tippling time. Additional paper work was a deal-breaker. Rex Johnson at College Station saw an opening. He wanted Fincher at the Argentine Pasos Street facility. The Kerrville dung beetle experiments had already moved to College Station with Richard Blume as scientist. Fincher and Blume brokered proposals for introducing dung beetles before the national program staff. The resultant meeting in New Orleans cast the die for a future that was exciting, devastatingly effective, but doomed by the chemical culture of the hour. Truman Fincher and

Richard Blume made their presentations, as did the entomology and parasite scientists mentioned in chapter 2.

Official Washington's reaction to what W.M. Bruce, Area Director of the USDA's Southern Region in Tifton, transmitted to staffers of the Dung Beetle Project is revealed in a letter sent to D.A. Lindquist, National Program staff scientist. It handled the problem, but it also contained subliminal hints that all was not well, that rationales were being constructed contesting a chemical approach with the secret lives of dung beetles…

"The Australian field work so far has failed to support the laboratory findings for the effectiveness of dung beetles in controlling the buffalo fly. No field data are available that indicate control of the horn or face fly in the United States. The major benefit of dung beetles, so far shown by field experiments, is the reduction of gastrointestinal parasitism. This aspect of dung beetle effect certainly should play a dominant role in considering future work, yet the report appears to be dominated by an unproven entomological viewpoint.

"It seems unwise to pinpoint a location for concentration of work in which only one of the proposed four thrusts can now be pursued successfully. College Station is suited for the importation of dung beetles because of the new quarantine facilities. The discovery of new beetles will have to be done overseas. The remaining two thrusts, ecology and biology, have been and are research objectives of the Tifton Laboratory which, with its collaborators, certainly is more knowledgeable and better prepared to carry out the work.

"It was brought out at the meeting that complete burial of feces takes place on the offshore Georgia island of Ossabaw, which is heavily overstocked with cattle and pigs, as well as wildlife. Recently, a similar situation was observed at Cumberland Island. These islands point to the fact that we have beetles available but we do not know enough about their biology and ecology to understand why the job is being done on the offshore islands and not on the mainland. Can these observations be continued from Texas?

"The foremost taxonomist on the Scarabaeidae is Dr. R.E. Woodruff of Gainesville, Florida. He is a collaborator of Dr. G.T. Fincher and has a great interest in the biology of dung beetles, especially in the associations on Cumberland Island."

The demurrers lined up like ducks as the broadside continued…

"A great variety of soils and climates can be found within 200 miles of Tifton. Similarly, big differences exist in the livestock industry within the same area. Certainly this is a fertile region for biological and ecological studies.

"In summary, I would like to point out that the foregoing facts were discussed at the meeting and further, it was shown that native species are available for burying feces on the offshore islands of Georgia. We do not know why they are successful there and not on the mainland. We would very well have the same difficulty with introduced species from Africa or elsewhere."

It was a harbinger of signal importance, and it would play itself out in the near three decades to follow.

Some 20 years after the semi-official start of the Agricultural Research Station program, Bob Steffen of Boys Town, Nebraska wrote to his senator about the dung beetle program. This prompted a USDA reply, a part of which is quoted here…

"For more than 20 years, USDA's primary research station… conducted research on the dung beetle as a means to control horn flies. However, it was determined that the beetles did little to reduce the fly population and this line of research was discontinued in late 1952. During the past several years, research at Argentine Pasos focused on the evaluation of the adverse effects of oral or topical chemical treatments of livestock on dung beetles and other… inhabitants of dung. The dung beetles were merely used as bio-assays for the chemical studies and not for dung management."

Beetle research was too difficult, the letter went on to say. Importation of beetles from all over the world would be necessary since no two or more beetle species could fill every area of the U.S. The yak-yak went on in carefully couched legalese telling the reader exactly what he already knew. In short, the program was being canceled as an economy measure.

The world takes strange turns, but life goes on. As a man drives over a country road in Georgia with a grandson in tow, the boy hears a crow making a racket in a nearby thicket. "What is it, Grandpa?" barely escapes the boy's lips before the bird flies up. "It's a crow," says the older man. He knows the next question before it comes, "What does it do?"

Questions from children are often profound. "Where does it come from?" and "Where does it go?" have stumped many a teacher. The boy's question about the crow caused the driver of the vehicle to find the right answer. Then a crow eating roadkill came into view. "This is the service that crows perform. They keep garbage from fouling the landscape."

In cornfields crows eat the larvae of corn borers. Longfellow's poem about *The Birds of Killingworth* comes to mind. The community killed all the birds and found itself inundated with insects that literally had the town's flora and peace of mind for lunch. The birds had to be brought back.

The man who was touring the boy through the countryside was Truman Fincher, an entomologist who loved crows, pigeons, starlings and sparrows. He was immeasurably fond of garter snakes, toads, salamanders and praying mantis. He saw beauty in cherries and the leaves of trees and the whole range of tree crops. Above all, he liked crawling things, snails, beetles of every stripe.

Down in Georgia, the man named Truman may have watched a scarab roll dung into a ball several times its own size. In classical lore, that dung ball was rolled over a patch of white sand until it became coated with shimmering particles. Now we understand how the Egyptians saw in the scarab a symbol of the sun god that brought life to earth. The beetle's egg held life like a jewel. The ball was buried soon enough. Now we ask, How much do we owe to Nature's insistence on balance and fecundity? How do we calculate the value of the service?

Let's take a trip to Egypt's land,
And sing about the beetle
Who rolls about on the sand.

Euell Gibbons once wrote the *Ballad of the Scarab*. You will read it with pleasure on the frontispiece of this book.

Phanaeus vindex, *a native dung beetle that figured prominently in the Truman Fincher experiments. It has metallic greens, is somewhat yellowish around the head, with a rough bronze texture on its "cape."*

7
The Research Drill

The message from the mountain is RRR (randomize, replicate and repeat) multiple times and SS PRP (statistically significant peer-review publication). Those with a somewhat suspicious nature spot the flaw in that sacred formula. Peer review has killed off more good scientists than Samson killed off with the jawbone of that other historic ass.

Bob Petit, a research professor at College Station, Texas spent a working lifetime studying humates and humic acids. Mandated into retirement, his work remained unpublished. From his Missouri farm he learned that his papers, now the property of the university, had not only been shelved, but destroyed. Much of what was done in early-day dung beetle research might have vanished except for a quirk in academia rules.

Truman Fincher left his Ph.D. dissertation behind when he left Tifton, Georgia for the uplands of College Station. The pattern he followed in his dissertation, *Ecological Studies of Dung Beetles: Scarabaeidae Serving as Intermediate Hosts for Nematode Parasites of Swine in South Georgia (with special reference to the beetles of the genus Phanaeus)* was simply ecological studies. The year was 1968. The humorist H. Allen Smith might have quipped that no one person could have written both the title and the thesis. That was precisely the case when I wrote my post-graduate thesis, and the credentialed guiding light of the work supplied my word-loaded title. Fincher defied the norm by writing a work so free of turgid language and impressive jargon, it may have contributed to the synopsis paragraphs that dozens, perhaps hundreds, of others came to follow.

Farmers and ranchers tend to say "scarab" when they are referring to a "dung beetle" because that's the term they all heard while growing up. It's the ancestral name, replaced now by the simple term "dung beetle," with the genus handy when peak interest sets in. Only rarely is a species name turned into a common name, such as is the case with the horned "taurus" or the poetic "gazella."

The most significant fount of knowledge comes to us through the auspices of observations made long before triple R and PRP became academic ritual. Answers have to follow questions, and questions rely on the insight and pragmatic observations from the living maturity of the scientist asking the question. By the time Truman Fincher as a young graduate came aboard the dung beetle train, it had already left the station. Breeding ecology had been established.

Here was an insect that rolled balls of manure, perpetuated its kind in a unique way, and gave a good account of itself by taking away a product offending the ambient air, thus helping soil organisms manufacture soluble forms of phosphorus, potassium and nitrogen. In addition, a cousin dung beetle creates tunnel residences for its kind and so aerates the soil that runoff is prevented even while manure is made to disappear with the finesse of a Houdini.

Many a lad attempting to send a dung pat flying like a hockey puck with a stick or kick finds the scatological puck nothing but a hollow shell, dung beetle residents having consumed its interior, leaving only a veneer of dung holding the puck together. Dung beetles that prefer bovine fare

are a particular lot. They like to slurp, much like a ten-year-old handling an icy drink in a 7-11 store. Certain types can move several times their own weight in a day, freshness preferred. Others chastise the insects, the face and house and horn flies that torment livestock. Dung beetles yielded more mysteries than questions before research really got underway in the 1960s and 1970s, most of which seems to have satisfied peer review, at least until toxic genetic chemicals got in the way. Questions about communities and populations were being answered with statistically significant data bases when the human actors in our story came on line in the early 1970s.

Fincher first set out to study the ecology of insects using dung resources and habitats with attention to the evolution of beetles not only in the South, but likely world-wide. Soon enough Truman Fincher discovered manure preferences that astounded many of his colleagues. That dung beetles evolved in harmony with the evolution of mammals was observed long before disciplined science and its peculiar codes of conduct got into the act. We speculate, of course, when we codify such an observation because our speculations lack addivity, meaning closely defined units of measurement and equal intervals with an absolute 0 as a base.

The variable of competition for dung, any dung, walks hand in hand with parasites encountered and the hostility of domestic species managing, or being managed by, the introduced diversity. It is the act of moving the dung resources to a safe location that prompts us to cite the tunnelers, the rollers and the come-lately dwellers. Each requires total scrutiny and juxtaposition to the excreta of choice, the battleground, the safe location, and the life cycle.

Rollers are on the job the minute a cache of dung hits the turf, soil or sand. Even camel drivers in the Sahara have to step lively in order to recover camel droppings near an oasis's concentration of dung beetles. Indeed, they have a sack and a scoop for immediate retrieval of the dung. In the dry air of the desert, camel dung rapidly desiccates and soon enough becomes fuel for a fire to make tea.

Ranchers with a lush grass sward look to tunnelers for their insect control and economic solution. Those underground passageways and chambers would be an engineering miracle if raised by power figures to a human endeavor. And the dwellers with their propensity for homesteading a cow pat also add their perspective to dung beetle ecology.

Early in the 1970s two workers counted 16,000 dung beetles gathering on the spot when over 3 pounds of elephant dung became available to them in East Africa. It took them barely two hours to roll away and bury that amount of dung.

In his college classroom opus Truman Fincher stated the problems of dung beetle culture, both answering and enlarging many of the points that other people in other rooms would revisit almost a decade later. There was a Phanaeus species in southern Georgia, and there were islands on which the culture of life and livestock hadn't changed much since the days of the great explorers, the Spaniards, the Chickasaw and finally the settlers. Fincher examined the life cycle of *Phanaeus vindex*. Fincher's learning tree was swine feces and the proposition that native dung beetles could match performance with the African gazella if! The "if" bounced that proposition back to the statistically significant requirement. There was the life cycle to be discovered and inserted in peer-review literature, and the case had to be made for overwintering, and he had his conclusion to offer that beetles could serve as intermediate hosts for nematode parasites. Evening flights of the much-studied beetle were clocked and presented in statistically meaningful arrays.

Did dung beetles transmit infections? Hellishly clever traps, accurate counting, and fresh swine-dung bait plus an experiment designed to comply with RRR prompted Fincher to add a new cast of beetle characters to the directory of those already discovered.

In less formal communications, Fincher was more apt to refer to the soil, the real wealth of the nation, and to the billions of unpaid microbial workers, which many farmers didn't want to admit had relevance now that NPK agriculture and rescue chemistry had arrived. It was hard to sell the farmer on the invisible, even though organic folks tried. In 1962, Rachel Carson called attention to the blunder underway in *Silent Spring*. After discussing soil and life, she asked and answered the central issue of our times. "We are rightly appalled by the genetic effect of radiation. How then can we remain so indifferent to the same effect produced by farm chemicals used widely in our environment?"

There are at least 22 sub-atomic particles that ride in on photons of light. Each has a name not to be found on the Mendeleyev Periodic Table of Elements, yet each affects human, plant and animal life in fragments so infinitesimal they might as well qualify for that mythical number "gazillion," or somewhere on the scale where words cease, but the numbers go on. The livestock in the soil does not do well in the presence of toxicity. Dung beetles in particular are sensitive to ionized chemicals. Some seem to prefer the dark, perhaps to avoid those sub-atomic particles. Farmers who are skeptical about the ability of microorganisms to make inorganic minerals organic for plant uptake are not quite as skeptical when they actually see dung beetles scarf away the nutrients that delicate human beings shun the way the devil is said to shun holy water.

"Dung beetles," advises Truman Fincher, "are easy to track and manage. They aerate the soil, usually delivering excess carbohydrates in the dung to microorganisms. Organic matter, so important for the carbon cycle and the nitrogen cycle, is built by dung beetles. Dung beetle tunnels soak up water like a sponge that is dry and thirsty." As for controlling pest insects, the man from Georgia says, "Gentlemen, hush!"

First things come first for those who take seriously, nay, even develop a passion for, land reclamation. The toxic materials have to go. No beetle can abide those man-made molecules. Systemic medicines for cattle have to be retired without ceremony, possibly incinerated. Such materials flowing through a bovine's intestinal tract are death to the beetle, to the colony, to the idea.

When brown beetles fly to cow dung as if directed by Nature's radar, the sight is much like a rainbow, refreshing, an I.Q.-inspiring experience. They do this all day, then again at dusk and also at night, as if to tone down those photons of light, those 22 sub-atomic particles. Few ranchers weigh out the "taking" of manure. Walt Davis, whom you met in the opening chapter, once measured that removal precisely. He found that his beetles, led by the ones that he and his neighbors call gazellas, removed a ton of manure per acre per day, this based on adequate rainfall and 25 animals per acre. A ton amounts to 90% of that day's manure. Walt's sharp-shooter can bite down into the soil up to 18 inches without finding a terminal point to the tunnels. The tunnels are the thing, especially when pastures are fertilized with ocean solids and only a dribble of nitrogen and even less phosphate — all there for the uptake, cafeteria style.

Truman Fincher sees the overarching simplicity of Nature's plan as the epitome of efficiency. Photons of light-carrying, sub-atomic particles arrive from the sun to stroke the needs of what Thomas A. Edison called God's greatest invention, "that blade of grass." Photosynthesis enables animals — large grazing animals like cattle, bison, water buffalo, horses, elephants, zebras, wildebeest, goats, sheep — to extract some of that solar energy and remand the rest to the elimination factor. But elimination is an individual matter and a universal impossibility. The largest of the dung beetles dispatched by Nature, now with human assistance, can measure over two inches in length. The smallest usually clock in at about an eighth of an inch in length. After that, all generalizations are false. The slurpers slurp, and the on-coming generation dines on solid matter. Somewhere in the mix is the lengthy food chain — Elaine Ingham would say "food web" — and all of it is dependent on energy from the sun.

With their antennae tilted so as to pick up wave lengths that researchers still dispute, the beetles fly and explore and settle in on their gold in the grass. How the chromosomes mix so that the beetle, of any species, can increase and multiply would make a book in itself, but the subject will be held to a modest paragraph here, to be elaborated later on.

We have a tendency to see insects in terms of ouselves. We are forever measuring social behavior, equating our taste buds to a presumed mechanism in dung beetles, and then registering abhorrence at what we see. Admittedly, some dung beetles exhibit recognizable methods of brood care. Nest guarding is viewed as cradle watching, and adulthood implies a rite of passage, such as purchase a pickup truck, all of the above a cartoon version of a life form in which the Y-chromosome is sometimes shattered or switched with the usual female XX.

There are beetles that exhibit advance behavior, with a proprietary interest in a brood ball, a natural directive. Nurturing the young, parental care, diversity unlimited, all these and more the researchers discover to their delight and chagrin. As with mammalian life parental care equals survival of offspring. And what about the role of mother? She can be excluded from the nest at a cost that most life forms are not willing to pay. Her role is so valuable that it asks for a downtrack chapter all its own.

We talk of predators and parasites, often forgetting that turnabout is fair play. Mother beetle may enforce birth control on occupants of the pat, but predators and parasites pay their lack of respect by preying on dung beetles. Ranchers who discover an interest in their newly-found cash cow do not ipso facto become entomologists. But they seem obligated to know something about the behavior of their servants beyond the simple knowledge that there are dwellers, rollers and tunnelers.

Some beetles construct a nest and hoard their dung fund directly under the pat. Others grab dung from the pat and construct that famous dung ball which they proceed to roll elsewhere. Still others homestead the pat, slurping the juices before desiccation takes over. And then there are those colorful scoundrels, the klepto-parasites, which so much remind us of career politicians. They steal dung, co-opt dung already sequestered. There are yet others, of course, and like that African inventory, they are too numerous to mention.

The research drill has addressed every conceivable nuance of dung beetledom, and still a moment's reflection conjures up still other topics for that degree-anchored paper. Absent too often is the real connection, not between beetles, parasites and dung removal, but between dung beetles and the owner's bottom line. The war of wits or instinct between producers and klepto-parasites suggests the unequal contest between pitchers and catchers in life's economic battle.

Withal, dung beetles really mean a countervailing force on the second side of the equals sign. Truman Fincher's opening paragraph of his University of Georgia dissertation wrapped a summary of the situation into a few whiplash lines. Breaking the life cycle at its weakest point became the focal point of his study and his life's work. The "weakest link in most cases was the free living stage," Fincher wrote. This was the starting point of the research drill.

Here are Fincher's paragraphs with enhanced identification for college workers and deletion of superfluous bows to the rules of college-paper writing:

"To a very large extent, the effectiveness of Nature's sanitation system is dependent upon coprophagous beetles. This is due to their habit of burying animal feces and carcasses. Few people realize the importance of the role of dung beetles in the ecosystem. Without them, our environment as we know it today would probably be very different, as suggested by the presence of CAFOs (Concentrated Animal Feeding Operations). The burying of feces by these beetles is an important part of the decomposition cycles of organic waste materials, particularly the nitrogen cycle. P. Gillard (1967) reported that in the absence of dung beetles with cattle on pastures in North Queensland, Australia, 80% of the nitrogen from the feces was lost by volatilization. When coprophagous beetles were present, this loss was reduced to only 5-15%.

"Bornemissza (1960) reported that due to the absence of adequate numbers of dung beetle species in Australia, cattle dung stays on top of the ground for years. In most countries where domesticated animals are indigenous, coprophagous beetles are present to bury the dung. Quick burial of the dung hastens decomposition, prevents loss of nutrients by volatilization, aerates the soil and increases the depth of soil containing abundant organic material. The prolonged presence of dung on the surface has an effect on the pasture somewhat like a noxious weed because cattle will not graze on the rank growth surrounding these dung pads. Taking the area under each dung pad and the unpalatable rank growth around each pad into consideration, Bornemissza (1960) reported that 5 cows would decrease the effective area of a pasture by one acre over a period of one year . . .

"The accessibility of the infective stages of parasitic worms to livestock is reduced when feces containing parasite eggs are buried. R.K. Reinecke (1960) stated that the recovery of parasite larvae from bovine fecal pats was greatly reduced when dung beetles attacked the pats in semiarid regions of South Africa. R.P. Bryan (1973) reported that the burying activity of the recently introduced Afro-Asian dung beetle, *Onthophagus gazella,* reduced the number of strongyle larvae migrating from fecal pats on pasture in Australia.

"Dung beetles in the southeast at the present time cannot adequately remove the ever-increasing number of cattle feces from pasture surfaces. However, in countries that have greater numbers of endemic species of grazing animals, many species of dung beetles have evolved that are capable of burying feces of native and introduced livestock within a few days after deposition. Gillard (1967) stated that dung beetles occur in profusion in the excreta of domestic and wild ungulates in South Africa and that the feces of these ungulates are buried by the beetles within a short period after deposition...

"The role of coprophagous beetles serving as intermediate hosts for certain parasites is also of economic importance due to the losses incurred from unthriftiness, morbidity or death of parasitized animals. Investigations of the clinical aspects of the dung beetle-nematode relationships of south Georgia have been underway for the past several years at the USDA Animal Disease and Parasite Research Laboratory at the Georgia Coastal Plain Experiment Station, Tifton, Georgia (T.B. Stewart and K.M. Kent, 1963).

"In 1967, Stewart and R. Davis conducted an investigation of the Acarina found associated with dung beetles in the Tifton area. Emphasis was placed on the occurrence and abundance of macrochelid species (phoretic mites), since great interest has been shown concerning their predation on housefly eggs and larvae (R.C. Axtell, 1963). To further elucidate the ecology of this swine-beetle-parasite relationship, an investigation of the coprophagous beetle species in the genus Phanaeus was undertaken at the Georgia Coastal Plain Experiment Station beginning in the summer of 1966. In conjunction with this investigation, a search for other Scarabaeidae beetles, which could serve as intermediate hosts for nematode parasites of swine, was also undertaken."

The official statement, made a matter of record during that era, is worth repeating...

"Because of a 'shortage' of native dung beetles, the introduction of carefully chosen species of dung beetles into the southeastern United States from countries that have similar climate and soil type seems to be a logical approach in helping to solve several problems now facing livestock production on pasture. A successful coprophagous fauna, however, must include various species which can live together with minimal competition, according to their ecological requirements and feeding behavior. Introduced beetles should be strictly pasture forms and coprophagous in both the larval and adult stages," so wrote Truman Fincher.

This much stated, and with due respect to RRR and SS PRP, we now move on to an in-depth look at dwellers, rollers and tunnelers.

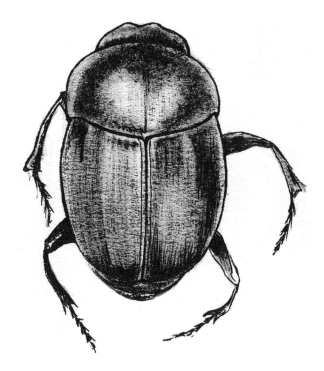

Canthon pilularius, *an American native, is smooth dark green with very little texture. It is one of Truman Fincher's treasured natives, albeit not a dweller.*

8
The Dwellers

When I was a youngster on the high plains of Kansas, barely old enough to be permitted to carry a .22 rifle to school for cottontail shooting on the way home, the pastures were full of discs. We called them cow chips. The sun dried them out in fast order, but usually not before flies and dung-loving insects had had their way. Beetles were sometimes evident along the creek or not far from a watering tank, and I remember seeing tumblebugs.

We — my brothers and sisters — picked up the thinnest of those discs and sailed them at each other like present-day frisbees. Someone had hammered in a nail high on the barn wall. The standing dare was to sail a cow chip so it would hang on the nail. My sole claim to fame before we left that farm during dustbowl days was success at sailing the disc so that it either stuck on the nail or to the wall.

Often we would find hollow cow patties, usually while gathering the dung chips for winter fuel. The stuff burned like charcoal in our gravity-fed furnace. Yes, hollow dung, and therein lies a story, if not a connection. It was rare to see dung beetles at all that far west of the isohyet line, that imaginary boundary from which rainfall decreases an inch a year for every 15 miles going westerly, or increases the same amount going east.

Ness County is the 5th county east of Colorado. I've been told that, one year it had less rainfall than Death Valley. Without moisture dung beetles do not thrive unless a flowing creek is nearby, or a slightly leaking water tank. They will also survive if they live within easy distance from a woodlot with soil dampness preserved by every form of organic matter and large-animal dung. As a practical matter, it is hardly possible to discuss any dung beetle in isolation from other dung beetles or insects, but we try. This is true, even though beetles vary in the resources they like.

The beetle species that once dogged the milling herds of buffalo seem to have vanished except for a few retrieved by laboratories. Those that have populated the high plains since then have done so very sparsely, and hardly any of these have proved equal to the task of survival during drought, nor do they seem capable of dealing with the dung gifted by cattle, the bison's replacement. Survival is as necessary to Nature as are photons from the sun. That is why high plains dung beetles are often micro-dwellers, partial to the mini-pellets of rodents and prairie dogs. Other species, weather permitting, favor sheep over cattle dung.

You can trail the several dozen species across Canada, through the Dakotas, into the mountains and plains of California, and tag each discovery with the name of the scientist who made the discovery, the name being carried with the italic of genus and species, until you have a veritable directory of dung beetle specialists from the beginning of century 20 to the present. The coasts, the South, the Texas rainbelt and hill country, all have their residents with their areas of operation and effectiveness as restricted as the territorial range of a coyote or mountain lion.

Those dung beetles near that Ness County watering tank that I mentioned earlier, were they dwellers or rollers? Seventy years later, I still have to say, "I don't know." Truman Fincher speculated that they might have been immigrants tagging along with those Hereford cattle imports that replaced both the bison and the Longhorns.

Suffice it to say that the beetles of most interest to the cowman are specialists, a few of Oriental origin (in Washington and Oregon), many out of European pre-scientific repopulation efforts, all of them almost mesmerizing to those who study them, less so for those who compute economic gain from dung beetle activity.

We are told that those glaciers of 10-12,000 years ago determined the survival rate of transplants and the survival and proliferation, or lack thereof, of several species, but not as much as the character of the dung resource. Early occupants of the North American mainland had only dogs as tame animals. Thus, we have those deer dung-loving beetles of New England, and the absence thereof where the deer and the antelope once played. Ecologists look for correlations when they chart the clearing of the forests during the westward expansion days of 1870-1890. Beetles of the forest found a pasture habitat unsuitable in the main, and therefore faded as the pastures grew larger.

The plains did not have trees to start with. Admittedly, the bison shredded its dung, allowing the low rainfall west of the isohyet to desiccate the manure. Through the 1890s-1930s, the dung beetle rated attention as little more than a curiosity among the wheat farmers who sometimes had a few head of cattle. If there were survivors, they faced annihilation the day that chemistry provided the grower with toxic genetic chemicals for internal and external use. The dung beetles called gazella apparently can't live very well above the southern end of Oklahoma, and the species that dwell above the Kansas line appear to have been natives, each now tagged with a Latinate name that would pepper each page with bothersome italic if presented here.

But what about the dwellers? They stay with a dung dropping until it becomes exhausted. They consume. They take the manure down under. And they are not terribly efficient. I walked those western Kansas pastures to and from school hunting cottontail, and I can't recall even one dropping disturbed by dung beetles except near the creek when it periodically flowed or near the stock-watering tank.

"No one imports dwellers," Truman Fincher told me. The reason is clear. They are not the workhorses of the family. Like baseball statistics, they have their glory deep in the bibliographies of scholars, but they mean little to the rancher who has tons of manure with a payload of nitrogen and phosphorus to be inserted into the soil. Hollowing out one of the 7-9 or more cow droppings a day hardly denotes a concentrated rotational pasture operation.

"The dwellers that we have," Truman Fincher told me, "came over with livestock on ships, mostly from Europe." Accordingly, most of the small dwellers are of European origin. Dwellers spend their entire life cycle inside the dung. These few facts stated, it now seems more likely that those hollowed-out discs resulted not from dwellers, but from actual burying dung beetles, whatever the stripe.

According to the literature, most work on dwellers has been done in northern Europe, especially in Norway. The rule of thumb is that the further north you go, the more dwellers you have. Indeed, dwellers are cold weather beetles.

Living dung beetle species are not alone in attracting the attention of academia, if not working ranchers. The Smithsonian Institute has a little booklet that presents an artist's conception of beetles found in the fossil record uncovered by paleontologists. Not much is known about the fauna of bison dung. Three dung beetles identified by California workers seem to have become extinct. It was speculated, if not found, that the disappearance of the three was due to reduction of the large-mammal dung supply. Survival seems to walk hand-in-hand with rollers and tunnelers.

Most dung beetle enthusiasts quietly ignore the dwellers. They are there, to be sure, but they are as unproductive as boxcar bums. They go to work in the spring, clear the pastures, then shut down as cold weather arrives. They have 24 hours a day to do their work. Fincher figures, depending on stock density, that they remove 2 lbs. of wet manure per acre per day.

Aphrodiines scarabs are small dung beetles. In terms of an inventory they make up the bulk of the cold weather dwellers. They eat dung. In fact, they live in the dung deposit without building any sort of nest or chamber. They live and die in place, and stay on the job even though tumblebugs collapse their home and roll it away in pieces, and tunnelers drag it down beneath them. Competition becomes a tempest in a dweller's dung platter, much less among rollers.

This bare outline should suffice for the reader, at least until he or she finds a good reading glass. On-scene magnification is bound to open an interest not easily comprehended by bibliophiles and dilettantes. Even so, the dweller is likely to be passed over by the lure of the tumblebug, the next entry in this brief four-chapter outline for the contents of the rest of this book.

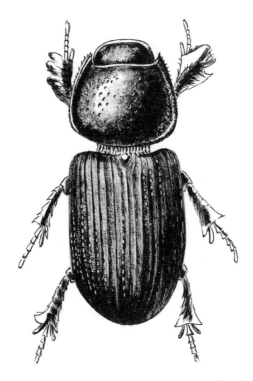

Aegialia rufescens *is yellowish to reddish-brown with a granulated head texture. It was first found in northern climes, Maine, Quebec and western Canada and identified near Marquette, Michigan. It survives well in cold climates.*

9
The Rollers

When Fletcher Sims, the compost master, fish specialist and conservationist, was growing up near Mexico, Missouri, the name for the resident dung beetles was tumblebugs. The most learned of the Sims tribe reached for scientific names, but relied on the abridged name "scarab" for the rollers in their midst. Fletch watched their activities and even compared their reverse rolling of dung to the locomotion of the female flea which only travels backwards. Backward travel wasn't restricted to the insect world.

Confinement feeders and businessmen joined the fray, always punctuating their progress by regularly jumping backwards.

That 40-acre pile of manure in and around Swisher and Randall counties in Texas feedlots served up fare no tumblebug could handle. It was bovine dung, all right, but it was as hard as paving stone. When Fletch set out to teach the nation the art of composting manure, cotton waste and materials all the way from corn stover to grass clippings and paunch manure, he recalled that sentimental memory of mankind, the scarab tumblebug.

A Howard rotovator performed only marginally in breaking down those feedlot slabs, so Fletcher Sims invented the Scarab Windrow Turner, a patent that was immediately infringed. It reduced particle size and performed as only a carbon copy of Nature can perform. It was a true scarab in its own way and helped undo the folly of greedy men and their bovine concentration camps.

But Fletcher Sims's scarab could hardly substitute for the real thing being colonized by the College Station breeders and propagators of beetles of Nature's own design. Still, rollers are not cherished in lush pastures. Their balls won't roll when impeded by the inventory of Texas grasses. Withal, rollers amuse people. In Egypt, they are as common as columns in the Temple of Karnak.

Like an ant farm with a glass window, the antics of the rollers positively amaze the casual viewer, intrigue the cow man as he calculates the manure removed from the pasture, and provide the scientist with enough variables to spawn a career. And yet, the scarab tumblebug is not a prized insect on the cattle ranch. Grass inhibits the athletic roll of the ball and badly compromises the burial process, ergo the brood ball. Still, any study of copulation, procreation, progeny preservation, and finally, release into the mature world of its year-long cycle fills more compendiums than do calculations of manure removal, the real purpose of this book.

We tend to think of sex in terms of the human expression, but Nature is not quite that simple. There is the matter of brood-ball construction, interment and the dual participation of male and female in a ritual that features less sexual pleasure than dedication to survival of the species. The architecture of underground quarters could easily detain us for several chapters. This side track will be brief, as well as deferred to the half dozen pages devoted chiefly to tumblers. There are too many types for instant comprehension.

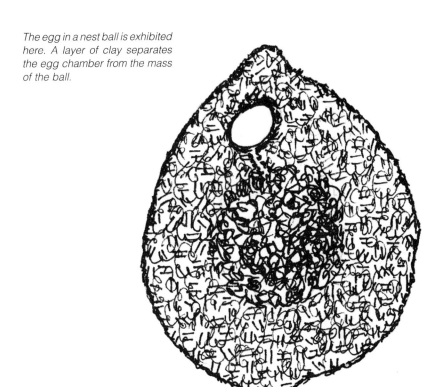

The egg in a nest ball is exhibited here. A layer of clay separates the egg chamber from the mass of the ball.

There are spherical and Italian sausage type "balls." Common to many is the fact that larvae use their own feces to construct the chamber occupied before pupation. The blueprints of DNA-guided construction meander in and out of genus and species, making all generalizations false. The cocoons, if such a word can be permitted, are edible and the egg chamber has its own center of gravity.

Onthophagus taurus comes to mind. Here, it is the mate's function to deliver, not the brood ball, but the fragments to be used in construction, like so many 2x4s. Some species do not rely on female cooperation, but there are others that feature cooperation faintly reminiscent of TV's *The Brady Bunch.*

The bland statement that brood ball construction is a female task has to be measured against the reality that thousands of species exist, dozens and hundreds in North America, even more in the tropics. Thus, the mechanisms of "brood ball constructed" and "an outer layer of soil added" have to be paired with a Latinate name that often seems to walk out ahead of the idea by a quarter of an hour. The minutiae attending each biography might attract more farmer attention were the dung beetles large enough to be studied without a reading or magnifying lens. Even so, as a Canadian geneticist put it to me, "Who's going to provide a grant for genetic research on dung beetles when a Senator Proxmire stands in the wings with a Golden Fleece Award?"

As it stands, we are forever indebted to hundreds if not thousands of entomologists for the more interesting sexual revelations that insect life has to offer. The much cherished, but lamented, Y-chromosome of the human male sometimes finds itself reversed with XX, thus directing the mix of genes in the progeny. The Australian agency CSIRO tells of dung beetles that merely roll fragments of marsupial pellets.

Scarabaeidae have their chamber habits. Onthophagus tunnelers seem to think less of their quarters than other genera that sequester their birthing place at the end of a remote chamber. Some cement their brood chamber with a female secretion that hardens into a smooth wallpaper, so to speak. Others submit their destiny to a slim covering of soil. Here, I am justified in saying "et cetera," meaning "all the rest."

Sexual reproduction is not for rollers alone, of course, but its anatomy seems to follow an evolutionary mandate answering the "survival of the fittest" norm, a term coined for Darwin by Herbert Spencer. Ball rollers compete for sexual favors. They fight like Turks and positively pursue mortal combat over a brood ball. As with Western society, the male, not the female, is lionized. It is the male that dominates the construction, rolling and interment procedures. As with certain human societies, neither sex is monopolized by the other. It is the dung on the ground that brings the sexes together to feed, to "carry on," and to obey Nature's injunction to increase and multiply.

Bi-sexual behavior and cooperation identify nesters. Every foible of mankind has its abstract counterpart in those hundreds and thousands of species: monogamy, polygamy, polyandry, even free-for-all sex, albeit under Nature's control. In any case, mere copulation is not a worthy objective in dung beetledom.

Protection of the young is a part of the reproduction program that has its own end, yet delivers benefits to the ecologically sound cowman never mentioned in the Cowboy Arithmetic or Cowboy Economics by the now defunct Oppenheimer gigantic cattle operator, then perceived to be a leader in the business. To have suggested to that giant that the humble little dung beetle was a major contributor to the bottom line would certainly have seemed absurd.

Competition for food is intense, this from coprophagous insects and rival beetles. Rollers spend a great deal of time rolling dung and therefore rate recognition denied their more secretive species. Predators like them for a meal. They do have glandular excretions that are to a beetle as powerful as a skunk's perfume is to a human being. This repellent factor saves beetles, albeit not when exposure outweighs defense.

As one Michigan cowman put it when the subject of dung beetles came up, "I have plenty of them."

"What kind? What genus and species?"

"I don't know, but they're tumblebugs. They navigate their manure balls between tufts of grass like a broken-field runner at a Green Bay Packers game. That's pretty good, going backwards!"

Those treasured balls yield up progeny in defiance of sub-Arctic cold that forever forbids interlopers from the sub-tropical south. Data for actual manure transport by tumblers are hard to come by because the crown prince and sovereign of Scarabaeidae dung movers are the tunnelers. Often without saying so, it is the tunneler that we call to defend the honorable craft of the scarab.

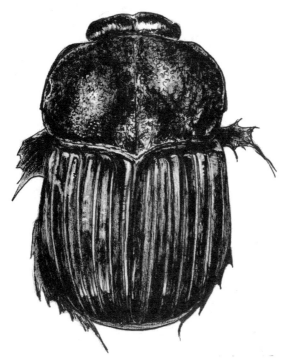

Onitis vanderkelleni *was imported in 1987 from Kenya via Hawaii and Australia as part of the College Station, Texas project. It has very black, deeply channeled wing covers and granular-textured forequarters.*

10
The Tunnelers

Recall if you will how Australian John Feehan takes school children into the field and gives them the water demonstration. It is a lesson worth putting into type and pasting on the bathroom mirror of every rancher in the country.

Proof is proof enough and Feehan makes the point that fencing creeks to keep animal pollution runoff away is merely a monument to the stupidity of man. The simple exposure to dung beetle architecture is an excellent prep course if only because it invites exposition and analysis of everything from the life cycle of tunnelers to building habits that might have taxed architects had not Nature shown the way.

Of all the beetle types, the tunneler is the most favored. The chambers and tunnels are constructed in advance of the event, much as war prisoners created their several tunnels before they launched their great escape in World War II. The hyperbole is used here quite deliberately because, gram for gram, the insect tunnels march hand in hand with the human effort measuring tunnel and size of the tunnelers against each other.

The tunnels serve several needs, starting with the investment in progeny, ending with fine-tuned calculations that morph into economic data revealed elsewhere in this book. The balls that characterize the beetle's DNA-guided role are built from the mass of material available. Some of this can be accumulated in the prime end of the waiting chamber.

Tunnelers seem to isolate their egg chambers from the food supply. Their underground pathways cross and criss-cross soils both vertically and horizontally. The lore of the brood ball has called upon researchers to get quite clinical. The female nurtures that ball and applies secretions that have antibiotic properties to ward off disease. As with the symbols of all fecundity, it is protection of new life that trumps all other considerations. The secretions to which attention has been directed cannot be dismissed as nutrient supplies.

They have personalities, these architects of escape fortresses and soil conservation. Burrowers have powerhouse legs. Their sexual prowess depends on select glands and anatomical markers that only a professor of taxonomy can discern. Well removed from pasture or forest, the professional worker tends to install findings into the literature that nudge the molecular, even atomic, plane of observation. For instance, the dung beetle — according to family, genus, and species — can be described as using its own feces, worked in a cephalo-caudal direction, to deliver a spherical or tubular cell. Walls are slicked down with a mucous secretion. The habitat for Scarabaeidae takes on the passion and dedication that anyone experiences when building a home or a nest.

Always preeminent is the protection of the larvae as they pupate on the path to development. The home may not be a castle, but it becomes a fortress. The cross-infection of chamber creation and behavior becomes self-evident when the species are studied side by side. The variations might detain us for a hundred pages were we to enlarge the subject to the fullest extent.

The beetle's role in the development of construction and architecture is so unique that it asks for a special paragraph all its own. Not only do certain species hang onto their primitive behavior, they also utilize the tubular chamber of earlier eons. The chamber is camouflaged by a crater lid. Analysis reveals that construction has taken place based on successive rings of excreta. Walls take on the consistency of veneer cement. The lid can be removed, which the adult can do, and does.

The sheer dedication of some Scarabaeidae to long "spousal" association may seem odd since sexual copulation is rarely observed by scientists. In one case, the female mates only once in a lifetime. If ball-rolling is more important than copulation among the rollers, then a countervailing pattern can be found among the tunnelers.

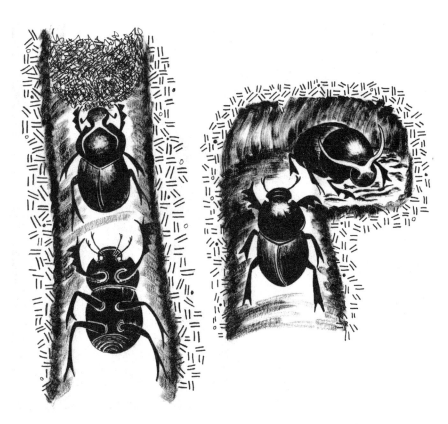

This illustration captures the next provisionary exercise of **Onthophagus taurus.** *Male and female cooperation is depicted.*

The tunnelers, for all their nuances and esoteric minutiae, nevertheless require us to recall the real reason for the cowman's interest, as opposed to the dedicated curiosity of the laboratory worker. The reason for being of tunnelers on pastures, as opposed to rollers, now comes into focus. In spite of our penchant for isolation of the individual, it must not be forgotten that diversity at the dung site with minimum competition is devoutly to be wished.

The above distills into an intellectual nectar easy to digest. Imported dung beetles should be pasture tunnelers because dwellers are always understaffed and rollers can't roll through the turf, whereas tunnelers most successfully aerate the soil. The mix is the thing. It serves to eradicate the parasites of both domestic animals and wildlife. The removal of feces decimates animal parasites, diseases and other insects. All of these things can be credited to the removal of dung from turf surfaces. Now we can see and measure the active fertilizer ingredients with resultant soil improvement, greater crop yields and improved forage.

Generalizations about burrowers or tunnelers are likely to degenerate into the types of conundrums made famous by medieval monks who debated the angel-carrying capacity of a pinhead. Make just one absolute statement and someone's likely to find an exception among the thousands of dung beetles now catalogued, with more coming in.

If Robert Ripley was still alive, he might well pick on peculiarities of dung beetles, particularly burrowers, and cartoon how they survive in sandy African soils by reinforcing tunnel walls with fresh excreta, an engineering feat repeated when man lined the Hudson tunnel and the England-France Chunnel with reinforced concrete. Unlike mad dogs and Englishmen, the tunnelers retreat when drought claims the kingdom for the season. Ripley would have to draw an illustration of those same beetles using dung pellets from one animal or another for nest-lining. Burrowing rodents often supply dung beetles with the wherewithal to find permission for life.

These few paragraphs are not intended to flesh out the rich topic that the name "tunnelers" implies. That chore has preceeded and will follow in a style best calculated to entertain and instruct the farmer and cowman — not least, young students hard on the hunt for a vocation. If successful, these few paragraphs may put a student or three on the road to becoming coleopterists, even engineers and builders. As I unfold the cast of characters, there will be side departures from the arena occupied by that single beetle family which, so far, has invited the lion's share of attention. Those other beetle families are there, like the proverbial mountain, and they are mentioned briefly, but as the narrative proceeds, the spotlight of interest will remain on the tunneler, first, last, and always — well, almost always!

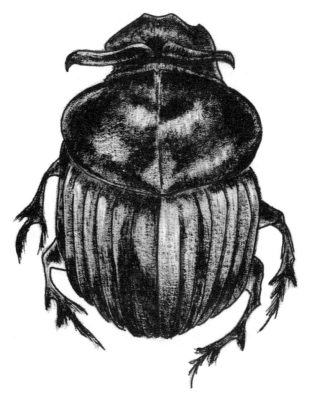

Onthophagus bonasus, *a 1980 Pakistan import, helped fill out the roster of 15 dung beetles that Truman Fincher and associates propagated and released. It is of russet color with some black overlaying the head and forequarters.*

11

The Klepto-Parasites

There is a fourth class of dung beetles, a class often omitted as if their propensities shamed the proud name of Scarabaeidae. Here, a new word enters our vocabulary: klepto-parasite, with or without a hyphen. The term is so exquisitely gorgeous, I can hardly pass it over, and no doubt will use it to denote some of the beautiful people that have turned the nation's capital into private fiefdoms. Perhaps the word will one day escape entomology and find application for creatures that inhabit Washington, D.C.

"Klepto-" means "thief," and parasitism has a meaning so well known even a child sees it operative in microorganisms, insects, mammals and human beings, every strata of society included. Klepto-parasitism is simply, for some, a way of life.

The term klepto-parasite was coined in 1869 by a coleopterist named Chapman. The type has been largely ignored ever since. The chief citation names the Aphodius genus. These rascals breed in the dung balls of others. The best example, however, is cited as *Aphodius porcus* parasitizing *Geotrupes* (a subfamily of Scarabaeidae) *stercorarius*. Here the klepto-parasite enters the egg cavity of brood balls, eating the eggs or otherwise destroying them, then laying its own eggs and converting the dung supply to its own use. Possibly, professionals will nix the idea of characterizing klepto-parasites as a type on par with dwellers, rollers and tunnelers, but with more than 7,000 species — with many habits unaccounted for — I can take refuge in the usual "as far as we know!"

As the name implies, these beetles reap where they have not sown. Klepto-parasites are not dwellers nor do they labor at tasks that earn for the family the respect of all who become aware of them. Klepto-parasites make no effort to build a nest or deal with the primary supply of dung on a work-a-day basis. They rely on the resources that other beetles have claimed as their own.

The term itself was validated in 1943 and has now entered the public domain. The dung thieves include names such as Aphodiinae and reach into the general Scarabaeidae family for recruits. All these Ponzi hustlers belong under the Scarabaeidae classification, offshoots from rollers or tunnelers. All klepto-characters are specialized genera. They have innumerable names no more difficult than thousands of other dung beetles, scientifically speaking, but no common names. These wayward beetles have the Onthophagus handle, but not the attributes of *Onthophagus gazella*, for instance. Some specialize in stealing dung from rollers. They seem mesmerized by dung balls being rolled. They follow behind zigzag style, like a second-rate gumshoe following a wayward husband or wife. When they reach an unguarded ball, they immediately have lunch.

Larger klepto-parasites rely on the availability of tunnelers to bring home the dung-world equivalent of bacon. They are a brazen lot, these klepto-creatures and they are as merciless as any parasite, insect or human. The klepto-parasites echo the philosophy of the professional hobo. They wait while a female excretes an exudate on a brood ball to prevent its dessication, then they furtively move in.

Both tunnelers and rollers invest heavily in their future progeny, permission for life that the klepto-parasites extinguish with alacrity. The klepto-parasites live as parasites in nests constructed by others.

Onthophagus gazella received their Linnaeus-type nomenclature because they like gazelle dung. They look a lot like brown June bugs. As dung buriers, they are nonpareil. They are survivors, hardy, prolific and in the absence of gazelle dung they gravitate to the bovine version. George Bornemissza suggested that their importation into Australia and their propensity for dung movement must have suggested their larders to klepto-parasites. Not that kleptos are that fussy. Like Yellow Kid Weil or Willie Sutton, who robbed banks because "that's where the money is," klepto-parasites go to where the balled-up manure is. They partake, then they parasitize.

Native species of dung beetles prefer native dung, a truism that at first governed the concept of repopulating the Australian continent. Bornemissza changed all that. In his native Hungary, dung beetles were up to the task. Himself a transplant to Australia, he reasoned that transplanted dung beetles would upgrade the scene the way that new blood enlivened both the American and Australian continents.

When *Onthophagus gazella* did well in Australia, it suggested itself for the U.S. southeast. It took hardly two decades to spread the species throughout Truman Fincher territory and some 700 miles into Mexico as well. The recognition of a new type, the klepto-parasite, followed. This is not to say that gazella enabled the evolution of the type, but the klepto is no simpleton. It wants easy pickins, and when it encounters gazella-packed dung, it's ship has come in.

Texas, the locus of high-level propagation, has an inventory of species considered enviable by others. Of the 90 or so species native to the U.S., less remain than first counted. Chemicals of organic synthesis have done their foul deed. As might be expected, the klepto-parasite type remains rare when other species populate their space in abundance.

Walt Davis, the Oklahoma rancher mentioned in the opening chapter, uses a stock density of 20,000 pounds per acre. Dung beetles have rolled up an enviable record keeping his 3,000 acres of pasture clean as a whistle, this from early June to September. He computed that his beetles were burying 2,000 pounds of dung per acre per day. That's a lot of manure, and one might assume that the few bites purloined by klepto-parasites wouldn't be missed. "Klepto-parasites?" asked Davis. "I don't think I've ever encountered them."

Perhaps the class does not deserve space with dwellers, rollers and tunnelers, but the name itself is like the song of the Lorelei, too sweet to be ignored, especially by the editorial writer who sees human klepto-parasites in high places.

Are klepto-parasites to be classed with those Scarabaeidae known as dwellers, rollers and tunnelers? The disagreement may not reach the level of dung beetle male and female having a set-to. Admittedly, a type of Luddite consequence and the overarching interest of the cowman clash, basic dung removal being the goal. Is the klepto-parasite a misfit in Nature? There was a time when carnival freaks were the only ticket for the study of genetics. Science has now graduated to the computer. Chromosomes can now be matched and mismatched on paper, genes can be reassigned, and the cell reevaluated in terms of DNA. I have no doubt that the klepto-parasites will receive such evaluation, much as human klepto-characters will eventually summon the interest of the psychologist, psychiatrist and geneticist.

The term klepto-parasite came into the English language as a tag for spider activity of some five families of spiders that in one sentence sound like an Italian opera. Still, words have a meaning, and "klepto-" means "to steal." Klepto-parasitism, stripped of flippancy, simply means feeding based on one animal converting food to its own use after another has gainfully earned the same. Stored food is part of the equation.

The insect world is full of good "for instances." Some bees deposit their eggs on pollen masses constructed by other bees. The term has expanded itself to cover theft of nesting material and other inanimate stuff by one animal to the consternation of another. The dung beetle klepto-parasite can target its own species or concentrate on any other handy species.

When it comes to kleptos, there is a paucity of literature, and even less interview material. "As far as we know," anoints every comment, speculation and hint for possible research. With attention to this phenomenon in deficit, many of these paragraphs can invite criticism and refutation, also on the basis of "as far as we know." But the name should endure because of its use as a metaphor and hyperbole, for surely animals that share a similar number of chromosomes have inserted in their primitive psyche something that finds development, albeit degenerative development, in some of the leadership that guides us all.

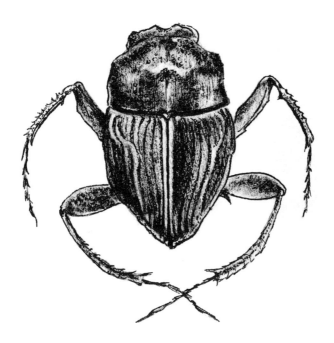

Sisyphus rubrus — *its classical name tells a great deal about this dung beetle. It was imported from South Africa via Australia in 1987. It too joined America's effort at repopulation until the program was canceled. It is rust-brown with traces of black. It has distinctive channels on its wing covers. The shell texture is slightly granulated.*

12
y-Sex

We can now recap the life cycle of the dung beetle in terms of a charcoal drawing, perhaps even in terms of the biologist. The terms of sexual reproduction are simple enough. Brood balls, hidden, guarded, nurtured, the apparent goal of many beetles, denote reproduction. Instead of going fast-forward, we would do better to go back.

There is a ranch in central Texas that touches three counties and might well be considered dung beetle heaven. Within hours after a cow patty has hit the turf, dung beetles on that ranch feed on cow platters, burying the rich, ripe materials in underground burrows, usually forming up a ball or rolling away a part of the deposit for interment. Few ranchers are cognizant of what is going on. In fact, it really doesn't matter too often nowadays because industrial agriculture has managed to annihilate populations of beetles that even a well-paid work force can't replace.

"Millions of hectares of pastures in the United States are lost to grazing each year because of dung accumulation and contamination." That statement anoints a report published by Truman Fincher, *The Potential Value of Dung Beetles in Pasture Eco-Systems.* Fincher's term "hectare" need not detain us. Multiply hectares times 2.47104 (or simply 2.5) to convert the mandated science of formal papers to usable farm English.

Dung beetles feed on the fluids of fresh dung. They use whole dung to feed their young. The scenario seems to be, lay the egg, nurture the young, go to work. As the songstress might say, "Is that all?" The answer is a qualified "Not quite." Dung beetles have been spared the rigors of live birth. The males have a phallus, as do sperm whales, but its dimension is measured in centimeters, not feet, ten feet in fact for the whale. Now we are forced to wander into that injudicious book, *The Double Helix,* to discover anew the DNA that choreographs life, dung beetles included.

For burrowers, sexual behavior is often defined by the longevity of the mating. Copulation is a hush-hush affair rarely observed by investigators. One species mates only once in a lifetime. Few Homo sapiens can identify with copulation that is so brief, as also is the courtship and mating behavior. By way of contrast, post-insemination becomes elaborate, cooperative, and cognizant of the investment in progeny.

Ball rollers have their own set of protocols. Copulation is frequent, if not entirely monogamous. In fact, the word promiscuity has no meaning for the species. Intromission is a part of the nesting ritual, but here again, the post-fecundizing relationship is cooperative, and ball rolling is almost always the work of a pair. The business of implanting the future of the species in a dung ball slicked over for hatchability of the egg is obviously more important than copulation.

Leigh Simmons, a professor in Western Australia, is an expert on sperm competition. Theory has it that the more sperm a male beetle delivers, the greater the chance for fertilization, ergo intromission. There are dung beetles that are as monogamous as a "walk the line" daddy in a 1930s movie. More likely, the dung beetle is as promiscuous as a herd bull. So, the professor speculates that the playboy types have larger testicles than the monogamous male. Theories are a dime a dozen: triple R, SS PRP proof is here. In fact, sex habits seem to be species specific. The bugs are either monogamous or otherwise.

When fresh dung hits the turf, the love triangle becomes operative. Picture the scene! The male keeps the female in the tunnel that takes form under the dung. This is the monogamous type. But in almost every species, there is the beetle Casanova, a real sneak. The moment the male guard is distracted, the sneak moves in like the Shadow of yesteryear. The Shadow quickly mates, then clears out. Thus, the professor's speculation: would the sneak rate in a sperm competition with a promiscuous stud?

It is probably unnecessary to point out that the guards are regular Man Mountain Dean types. They fight for females and guard them jealously. Horned beetles are of this type. Leigh Simmons has now revealed that the sneak has large testicles, meaning that he invests more resources in his avocation than the upright citizen that plays the game according to our version of Nature's Hoyle. Sneaks are puny fellows compared to the monogamous straight-arrows of beetledom. Sexual reproduction seems to pair with social life: insect, larger animals, or human beings. Such an inves-

An artist's rendition of a typical scene when impregnation takes place.

tigation may seem arcane, but the business of a pint-sized beetle providing one end of fecundity better than the chest-thumping — if beetles could thump their chests — giant horned guard merits reflection. The Australian professor goes on to make a human male extrapolation, but I'm not going there because there is still a lot of the dung beetle story to be told.

The scholars who have made their SS PRP comments available to the public tell of scarabs whose secondary sexual behavior is less developed, and of some whose sexual display is absent or minimal. Ball rollers do not avoid same-sex contact. Not that life is peaceful. Often there are challenges regarding a ball. Resultant combat is as wild as any sword fight in an Errol Flynn movie.

We tend to think of 23 pairs of chromosomes for the human being, an extra Y having gifted or cursed the male form, give or take, thus the battle of the sexes, XY vs. XX. No such rules govern the insect world. The spider accepts the world that makes spiders, then promptly kills its provider. That crawly creature might well adopt the words of Gilbert and Sullivan's Princess Ida, in which the male animal was characterized as Nature's sole mistake.

Birds turn upside down the XY-XX arrangement that humans find dominant. Here, males are XX and females are XY. What then are the rules for the object of this inquiry? Hundreds, if not thousands, of workers have spent their scholastic years finding out. It is now difficult to find any area of inquiry that has escaped the scrutiny of researchers. In this narrative I will attempt to recite basic questions and answers. But first, there is that trail out of Africa for life itself. Some people accept the evolutionary path charted by ancient Greeks, then by Erasmus Darwin and his grandson Charles. Others, such as geneticist Mae-Wan Ho, see the pure theory as a fairy tale for grownups, one "that pretends to answer everything and answers nothing."

I won't go there in telling the dung beetle story. If it is true, then men, worms, flies, even dung beetles have had their DNAs decoded. That story is most challenging, but the narrative here has to do with that missing new accounting principle, possibly one that sees the overarching principle as the smallest component. Such a system allows farmers and ranchers to consult the dung beetle for assistance with the goal in mind of profit for the cowman.

Symbols of males as the first sex are decreed by the Y-chromosome, although dominance has to be questioned. Some investigators warn that generalizations based on group and/or other considerations must be dismissed, remanding the study to individual conduct as shaped by the evolutionary process. We are at a loss to explain the low sexuality of the tunnelers. All mammals and not a few insects start life as females. The human species, of course, has nipples to prove it.

Reproduction for Scarabaeidae is quite low. The coprophagous species almost always has one female in charge of one single nest with one single egg. A higher fecundity exists, of course, but the female with 180-200 eggs is rare, even under laboratory conditions. The best, as a general rule, is 20 eggs. The size of the female is also relevant. Each offspring takes its toll. That is why great numbers are achieved only in the laboratory. The time and energy required for nesting cause rapid deterioration of the female. Brood care is a function of sex, perhaps instinct. Nature causes the young to survive and adults to spend themselves on behalf of their progeny. Competition and environment must be factored in, but nevertheless, progeny are sheltered, catered to by parents, and nurtured until total emergence. The nest is an insurance policy for offspring survival. Parents know this, and invest heavily, time, energy, and resources.

Cleaning of nests is a directed verdict from Nature when scarab progeny are expected. It does not stop there. Division of labor between the sexes has been ordered by their DNA, as is feeding of the young. This cooperation in nest building and stockpiling dung for food may be a short-lived phase, but competition at the cowpie is intense, each foe being a parasite or a rival beetle. The battle is always over the business of sequestering a valued resource. Chamber storage thus achieves the status of a larder.

Scarab dung beetles are chiefly generalists. True, they search out their preferred dung in terms of age and consistency, and thereby demolish insect egalitarians. But specialists invite our attention and the attention of cowmen. For reasons too diverse to catalog, the natives seem not to have developed a connoisseur's taste for bovine excreta, this because the continent's native fauna were different even 200 years ago. Then came the sudden — in terms of geological time — appearance of Bos taurus and Bos indicus on the continent's grasslands. The Onthophagus species in Australia is a case in point. Beetle experts tell us that as many as 175 of these beetles have been collected tending the feces of one wallaby.

The brood ball taken down under and provided with a single egg is legendary. Burrowers have a nesting system all their own. Here we turn to the artist's pen because a picture in this case is certainly worth the proverbial thousand words (see page 91). Some Onthophagus subsist on bat guano, but no one seems to know how they nest.

All assembled, the techniques of brood ball construction and maintenance, along with the pertinent birthing process, would make a book the size of this one, and perhaps detain us from our chief objective, namely, the clarification of the primary reason for dung beetle repopulation, which is to assist the livestock industry.

Suffice it to say that researchers have been unable to find common denominators except in providing and preserving brood balls, but one absolute survives. Scarabaeidae are quite superior as competitors, ecologists and providers of economic benefits to the rancher. Their nests are superior in the opinion of many, and they stand alone in the insect world as prime candidates for emigration, assisted or otherwise.

Television is never done with sex, and coleopterists would never be done with it either if not for the fact that they have to move on. It is probably enough to say that almost all female beetles have ovarioles, or small compartments, in the ovary, the more the ovarioles, the greater the egg production. The reproductive potential of many beetles outruns the cloning procedure in any case, which may or may not entitle the klepto-beetle to a natural mistake status.

It is safe to say that some species have freedom of choice. They warehouse sperm in a special sac, then choose which egg to bathe in the life-fluid as it passes. There are hooks and spines to enable coupling and intromission when these most numerous animals on planet Earth obey the order to reproduce. Like the turtle, the dung beetle has a heart, several feet and widely separated pumps that obey a coordination signal seldom duplicated in our mechanized world.

It is disconcerting to know that human marital disputes are not alone in spawning violent outbursts, abortion and infanticide. Brood care takes a back seat when the explosion occurs. It may be that male chauvinists assign blame to the female. The tantrum results in an attack on eggs. The end scene has the male or female destroying the eggs in brood balls. Both abandon the nest if a brood ball survives. This scene is repeated with or without cameras running. Some species enjoy nuptial feasts. Some destroy brood balls and nests as a form of birth control. It seems that a female's work is never done, the male — well!

Many beetles are switch hitters. They prefer dung, but they also treasure carrion and seem related in spirit to characters such as coffin beetles and working giants such as the American burying beetle, even though of a different family.

These basics leave unanswered the kind of probes that have made human genetics so fascinating. How, indeed, does the insect world determine sex? Is there a Y-chromosome that stamps its image on the tribe or on the species? Must we return to Gregor Mendel, or to that 1902 discovery of the cellular difference between male and female, really not too long after Mendel made a matter of record his findings regarding peas in terms of heredity? Malcolm X scrapped his surname, Little, in favor of the single letter X. We can do no less when we scrap most folklore baggage for the letter Y.

It is the Y-chromosome and its DNA associate that tantalize that nemesis of criminals, that friend of genealogists, perhaps the last link in understanding how it is that certain species of dung beetles bury more dung than their amateur counterparts. It is not the goal here to tell ranchers what they already know. For this reason, and for reasons left unstated, there will be no retreat into the esoteric language of high science.

It is said of the human being that sex and the lie are partners in the same bed. This is not true of the insect where scrutiny proceeds under a magnifying glass. Almost everyone knows that the Y-chromosome converts the human embryo from its female beginning to male. The geneticists tell us that the human male is being deconstructed, first by estrogen-testosterone mismatch, then by homosexuality and baldness.

Y-sex leaves unanswered those rare beetle specimens that produce beetle populations without the assistance of males. Single-sex reproduction, often belonging to Greek mythology or residents of ancient Lesbos, is a small fact with all the hoopla and few of the deficits of Dolly-the-sheep. The female capable of conceiving is rare, even if efficient in the eyes of bean counters. Sexless marriage is usually another name for sterility, still the clone has been observed even when under assault of errors in DNA that might be cancelled out by genetic material from the Y-chromosome. It works for some beetles even if the experiment holds little hope when tried as a way to reproduce virile bulls.

Oddly, other life forms construct their males quite differently. Genes are used as are chromosomes, whereas others have a capability of redirecting sex when certain stimuli show up. The male direction guided by Y is no absolute, at least not in the insect world. Billions of years have figured in the dung beetle's making of sperm.

Certain bedbugs receive sperm through their body walls. This is followed by the homosexual rape of the copulator who then passes the potent material on to the next available partner. Sperm transfers genes, but it also transports fragments of DNA, this over eons of time. Slime mold is said to have 13 sexes, 12 of them male-ish. If Nature has supplied special erectile chemicals to the dung beetle as an organic Viagra, it has escaped the attention of this scribe.

Wasps have a control that is hard to understand. The mother can choose to lay either fertilized female eggs or unfertilized male eggs. Some parasitic types park their eggs in caterpillars. Selection of son or daughter depends on size, a son for the smaller host, a daughter for the larger ones. The rules for fecundation change as the quiz show moves to new territory. Often, sons are expendable because of the fecundity factor.

Honey bees have 16 pairs of chromosomes, but these cannot be labeled sex chromosomes. They have a haplo-diploid arrangement, meaning that females are the product of the union of sperm and egg. The males generally are the product of an unfertile egg. With sexual reproduction a given for dung beetles — "as far as we know" — a clinical question asks for an answer. What is the chromosomal package for each sex? An answer has to be provisional because of the 7,000+ species in the beetle family Scarabaeidae, only about 26 of which have been investigated.

Aphids have a deserved reputation for a pseudo-cloning procedure called parthenogenesis, or sexless reproduction. Even so, the insect requires sexual service at least annually. But for dung beetles, chastity is worse than a bad end. It is the blood brother of celibacy, the end of the one natural system that really works.

There is a reason for this, and it is probably the same reason why various plants with asexual reproduction fade and perish quite early. With a male sex, genetic damage accumulates and more perfect phyllums take over, either this or insect predators win out. Here, the agronomist speaks up, Yes, this is true. The great hunger of 1845-1850 in Ireland was a consequence of potatoes vanishing under fungal assault. It was a sexless variety. Instead of chastity, chastisement attended the sexless spuds, a lesson that durable dung beetles may have noted.

This much stated, it is nevertheless a fact that the sexless banana has escaped the fate of that Irish potato variety — so far! Sex, to be sure, is the only escape from an evolutionary dead-end. When the Y-chromosome is absent, the mutations created by life's wear and tear cannot be shed. As a consequence, junk DNA claims homestead rights and finally can't be given the gate.

Parthenogenesis is really a reproduction arrangement, albeit not a clone. The males are all different, not the identical twin copy expected of a clone. The female warehouses sperm in her spermatheca, which is a sperm storage organ. When she lays an egg, she can decide — at her discretion — male or female status for the hatch. The size of the cell seems to govern the decision. Drone or worker seems to relate to the chamber available for reproduction. Instead of a sex chromosome, they have a sex locus, meaning a single gene. The forms are many. There's a locus for eye color: blue, green, brown, etc. There are about 15 different sex locus genes.

There are about 20,000 species of bees. Do they all have different sex-determining mechanisms? "Not really known," say Cornell University bee specialists. I recite these few details because systematists find dung beetle research an incomplete pursuit without reference to different orders, families, genera and species of the beetle phylum Arthropoda. We are hinting at the possibilities for sex determination in a world less known than the fabled canals on Mars or the tale of Atlantis. This much stated, we nevertheless expect sexual reproduction to be accompanied by packaged chromosomes. The chromosome numbers stacked up so far range from 10 to 18, this for the 26 scarab species studied.

The sex-determining system varies. Some species have neo-XY, and others have XYp. The small p confuses lay readers until it is revealed that p stands for parachute. This means that there is an extra floating chromosome which looks like a parachute opened. No coleopterist would care to suggest that Y-sex is a given. Parthenogenesis has its role in aphid reproduction, but we cannot expect such a role for dung beetle reproduction when hardly any of the 7,000+ dung beetle species have been studied. The above information is nebulous stuff, certainly nothing a cowman might wish to follow. Scientific probes in this area may or may not involve the dung beetles of our acquaintance, but merely hint at the hidden lessons in the unopened book called Coleoptery.

Definitive answers regarding Y-sex, the chromosomes involved, and the procreation details for well over 7,000 species cannot submit to a meaningful common denominator. There was a time when all dung beetles were classed as Scarabaeidae. Quite recently, additional groupings have been assigned. Even so, almost all Scarabaeidae have 2N = 20 chromosomes with a modified XYp parachute as expressed above. It appears that the Y parts with two X-chromosomes rather than a single X.

There are other sex systems for Scarabaeidae. One species has 2N = 20 in the female and one less (XO)2N = 19 in the male. A few species have high counts: 2N = 48, and 2N = 28. These last have a true XX arangement for females and the well-known XY package for males, as is the case with human beings. We may wonder that enlarged lower case "y" stands for male. Why not "m" or some other letter, as is the case with some insects that seem to reject all or part of the Y-chromosome scene?

At least one or two of my associates in an army unit made up of 18-year-olds during the "late unpleasantness" with Germany and Japan believed they had the answer. That lower case "y" has two arms, the long arm and the short arm. We all stood a short arm inspection now and then. The order for the formation was "caps, raincoats, and boots only." A long line would snail its way to the inspection station. Upon arrival, each soldier would throw open his coat flasher style, and the physician made the short arm inspection.

In the "Y" symbol, it is the short arm that decrees the male sex of the progeny. The human male has a YX arrangement at the apex of the 23 chromosome count, to pass on to male offspring, not female. The agency of that transfer is the short arm. Fact or fiction regarding the naming of that memorable inspection, I never found out. But the medic swore by the story. There is no comparable drill for beetles — as far as we know!

All of the universities contacted made the point that Nature does not clone. She practices parthenogenesis. Cloning was ratified by headlines in the case of Dolly the sheep. It involves an egg with nucleus removed and nuclear transfer of an uncommitted stem cell with its entire chromosomal package in tow. The Petri dish spawned life that can be transferred to a uterus and even grown to term, ostensibly to harvest mature stem cells for therapeutic use. This lore isn't a part of the dung beetle story but is included here to illustrate a sex system that Nature has not tried. The Missouri amendment on embryo stem-cell research has spawned claims of bad science and sophisticated barbarism leading to eugenics. *Newsweek* magazine left in escrow a comment likely to be recalled down-track as the debate becomes national. As the medical movement toward growing humans for valuable spare parts for the benefit of still other humans gathers speed, deep-seated moral questions can be expected. If perfection of human beings arrives, the specter of eugenics is bound to remind us of the mid-century 20 use made by a civilization turned upside-down.

Is there a masculine moment for insects? Someone somewhere may have an answer, but the behavior of dung beetles suggests not! The dung beetle uses the female egg to copy its own DNA. Thus, genes move and scramble like a fighter squadron, otherwise the copy of the female would be no more than a Dolly-the-sheep clone. Fertilization is a requirement in Nature, even if a deviation allows virgin birth in some few insects and leaves hermaphrodites to service and incubate the young in other cases, but these are anomalies that Nature seems to have spared the central characters in our story. Sperm is treasured, hoarded and so powerful that it inserts new heredity as each cycle of life adds generation after generation to the genetic mix.

Tactics used by dung beetles to attract a mate put to shame the dating game of some fauna, but fall quite short of the antics of the peacock and exotic residents of jungle canopies. Bark beetles issue a shrill sound. Other tribes in other environments bang their heads, drum their abdomens, and test the scale of notes excused from opera because they range too high for the human ear. The flashing light of the firefly is in a class by itself. The outgoing signals are as little strobe lights. The female "come-on" signal is a solid light lasting longer than a flash. Dung beetles are less flamboyant but their rituals are as encased in tradition as the wedding feast in any human ethnic group.

There are beetles that reveal themselves as one of two types, one with an enlarged version of a certain chromosome, the second with a smaller version of the Y. There are men who lack a discernable Y-chromosome and are possessed of two copies of the X. Thus, we now need to consider the genetic accident. A broken Y is implicated, having become attached to the X. Eggs fertilized by an X are female in any animal. It is the Y that delivers the offspring into maleness. The sex-determining region of the Y, geneticists tell us, was found in 1990, the SRY. SRY can interdict the run to female status.

Sperm and egg each contain half of the donor's DNA. Withal, sex is the great connect. It presides over new blends and ultimately the rise of one species over another. It drops into the out-basket of damaged DNA. Sins of previous generations are preserved simultaneously. In short, the interest of the Y is maleness. In any species, it is the Y that prevents the female from issuing clones. The pattern should by now be familiar. Even in dung beetles, it is the female that makes the largest investment in reproduction.

Parasites evolve with the skill of a virus. They steal identity or invent a new one. The insect world uses Nature's capital in ways that baffle, even contradict, but also confirm the reigning pattern of all life. The cell shuts down or can shove off, but the female beetle has to make the decision, it seems to us, to remain with the potential progeny. Like a herd bull, certain males seem hell-bent on making a *Guinness World Records* entry in how many females can be inseminated, yet monogamy is also evident in certain species.

Antagonisms survive like an Irish grudge and resultant battles recall mortal combat in Korea, Vietnam or Iraq. We can impute anger, rivalry, even revenge to some of those dung beetle genera and species. Still, nothing can compare to the destruction wrought by man. Always, man shows the way. Even myth confirms this.

Pasiphae, the wife of Crete's King Minos, gave birth to the half-man, half-bull Minotaur. Totally chagrined by her husband's debauchery, she brought retribution to his mistress in a manner never duplicated in all of Greek mythology. She invented necromancy and caused Minos to inseminate his mistress with sperm that brought forth snakes and venomous insects which devoured her intestines. Alchemy from the Devil's pantry seems to have done the same thing, figuratively speaking, to the dung beetle. The Minotaur's mother was the progenitor of the industry we call chemical.

Semen is the fuel that drives life forward for most animal and insect forms. It is a gift, but it is also a tax. Insect sperm is like wine at a Borgia feast. Some aspects suggest digestive enzymes. Some chemicals close down the reproductive tract, cancel out the drive to mate and command the production of multiple eggs. As we have seen, some males demand monogamy while they themselves pursue a playboy role. Some insects, like the horse fly, settle for a one-night stand, to borrow the human idiom. Afterward the female's mating days are over.

There are perhaps a hundred special proteins in insect sperm. They speak of attack and defense. Much the same is true of animals. Insects seem unafraid of the risky liquid. We must not assume that the rules are always the same, insect to animal to man. Some mice achieve maleness without a Y-chromosome. The details do not belong in the rural grammar, except for true seekers. Suffice it to say that some 2,000 proteins have been sequenced end to end for mice and men, and only one in ten differ. They service reproduction, the immune system and internal medicine. The Y of which we speak harks back to before the crash of that meteorite in the Yucatan Peninsula. We see dung beetles die out, yet we see them trace their ancestry back to the Pliocene era.

Now we see that the brood ball of the tunneler or roller is merely a minor part of the story. The well-fed beetle presides over fecundity. Dry summers, cold springs and toxic seasons of any stripe interdict population expansion, ergo these things offend the dung removal mission.

There are tomes on nesting that would make a *Fortune* magazine study on human housing pale into insignificance. Soil is the timber, the insulation and the fortress. Site selection reveals a natural instinct denied house buyers who never consider noise levels, morning and evening sun, or drainage. These are built into the beetle's DNA, its chromosomes, genes, and cells. Some researchers have even computed the investment made by the female, the offsetting cost stated as reduced fecundity.

Nests have been classified thusly, 1. Construct ball from food source, 2. Roll to a site, 3. Bury with single egg.

Bury and consume is the name of the game.

The observed birthing ritual relies on an umbilical of brood balls. The trail to and from the manure pile remains unobserved, at least for that most courted of dung beetles, the tunnelers. Sexual behavior is seldom priority stuff when schoolmen record the birth and death cycle because almost all attention is focused on the brood ball, its special chamber in the underground maze and the role it plays in pre- and post-emergence nurturing.

Male and female dung beetles of the burrower type appear to be real egalitarians, sharing the brood-ball maintenance work. The ball itself can grow to two inches in diameter and the egg's position in it is part of Nature's design. This design decrees residence in the small end and armor-plating made of clay. Is the egg forgotten while male and female beetles march off to a new drummer? Not so. The partnership requires total interest, or the egg won't hatch.

There are tunnels and more tunnels, feeding tunnels, main tunnels, lateral tunnels, tunnels that seem to have taught man his tunneling skills through the Alps, under the East River estuary, in mines and even under the sea.

Rollers, having performed their sexual duties, manage their birthing chore somewhat differently, yet somewhat the same. They select their interment site with the insight of Hawkeye scouting a campground and they roll their charge to it, always powering the burden backwards. Topsoil beckons and clay often presents a barrier. When a larva escapes from the ball, it taps the fecal mass. There are beetles that live for two years and lay up to 34 eggs each season.

Bisexual behavior among beetles for short and even long periods and construction of more than one nest are anomalies of dungdom. Indeed, such pairs seem oblivious to the cost of parental care. Species and sex recognition do not rely on direct contact in many cases. Both can be achieved at short range with bartered pheromones. Sometimes a pair's members touch each other in the manner of a kiss. There are sexual differences in numbers and locations of glands likely to secrete pheromones. Some glands are positioned in the mandibles.

Instincts that govern sperm donation extend themselves into the nesting ritual, the sourcing of food, the establishment of the pair bond, supplying the nest with dung, forming the brood ball and protecting new life, though not always. For some, the nesting season commands a nest every few days. The Sisyphus copulate before each nesting season, gaining enough seeds of life for the year, but others present and receive semen once in a lifetime.

All Scarabaeidae have only a single ovary, a factor in fecundity. Smaller species that seem undersexed have an extremely low propagation rate. A general principle emerges. As long as the female tends to her brood ball, ovarial development is inhibited and birth control is realized. Much the same is true of the male caught up in the business of parental care.

We are rightly amused by the dating game, this business of the male being "hard to get," and the female with the dung beetle version of the proverbial sick headache. Perhaps a rewinding of Nature's procedure can help us understand subtleties that are never a part of Google search-engine findings or brochure fare when dung beetles are sold for colonization. Assignment of nest details would require a biography per species, literally, and is well beyond the purview of a handbook for laymen, farmers and ranchers.

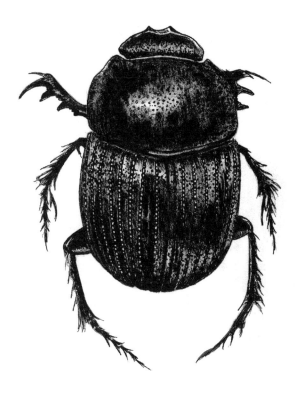

Onthophagus depressus *was found in Georgia already in 1937. It is charcoal gray to black, a perfect soil camouflage.*

13
Cast of Characters

Now that we have surveyed the general scene, the time has come to identify in a realistic way the cast of characters of which we speak. The task is not simple, as I intend to make clear with fragments of what journalists call standing copy. A comparison of weeds to dung beetles may be justified as these passages from my *Weeds: Control Without Poisons* and a few interruptions suggest.

The grammar of the subject requires that we learn a few new terms. I state these now and defer a longer, more detailed explanation at a more logical stopping point in this narrative. I start with phylum and its name: Arthropoda. The class covering dung beetles is Insecta. The order is Coleoptera, and the specialist in the subject is called a coleopterist. Under class comes the family: Scarabaeidae. Now we come to genus and an assortment of names that could fill much of this book. The chief genus discussed here has been Onthophagus, species gazella. It is this story that Nature's bard tells with all the drama, pathos and emotion of Homer himself.

That single common name for a family is not entirely unique. Of the several hundred families scrutinized, only a few have a layman's moniker. We all know about the boll weevil. It likes the cotton plant and remains the first great illegal invader from the South. It crossed the Rio Grande at Brownsville, Texas in the early 20th century and claimed the southern tier of states — plus Arizona and California — as its own almost as fast as cotton crossed the Mississippi to become a Texas staple.

A few of the beetle families have common names not unlike the sobriquet "dung beetle" that lay usage has bestowed on the Scarabaeidae family. An incomplete list includes telephone skiff-pole beetles, minute bog beetles, whirligig beetles, crawling water beetles, burrowing water beetles, trout stream beetles, predaceous diving beetles, wrinkled bark beetles, ground beetles, water scavengers, false clown beetles, feather-winged beetles, carrion beetles, rove beetles, ant-like stone beetles, stag beetles, bedbugs, hide beetles, rain beetles, earth-boring dung beetles (subfamily Geotrupidae), dung beetles, June beetles, May beetles, Japanese beetles, chafers (all members of the Scarabaeidae), jewel or metallic beetles, soft-bodied plant beetles, Texas beetles, long-lipped beetles, glow worms, soldier beetles, wounded tree beetles, powder-post beetles, death-watch beetles, ship-timber beetles, soft-winged flour beetles, dry fungus beetles, sap beetles, flat bark beetles, shining flour beetles, silken fungus beetles, lizard beetles, fruitworm beetles, handsome fungus beetles, ladybird beetles, tumbling flour beetles, cylindrical bark beetles, wedge-shaped beetles, blister beetles, false blister beetles, fire-colored beetles, narrow-waisted beetles, bark beetles, ant-like flower beetles, long-horned beetles, leaf beetles and leaf-rolling beetles.

This partial list merely hints at the complexity of the beetle world in terms of families, with the majority without a common family name. As for children, roll call would require a compendium as big as the Manhattan telephone directory.

Weeds, for instance, have two naming systems, one bestowed by indigenous peoples and come-lately usurpers and a second named by scientists. The latter's venture into the naming game is called taxonomy. This more perfect procedure is simply a filing system for biology. The common names invoked by dung beetle aficionados are hardly a half dozen, whereas the scientific roster includes thousands of well-crafted Latinate syllables stitched to still other syllables, often including the name of the beetle discoverer's surname identifiable by the suffix-tag "-i."

Those who seek to trace the evolutionary tactics of the tribes and sub-tribes have their work cut out for them. Even a cursory look at a challenge that makes Job's patience seem inadequate suggests going where Star Trek has never gone before — into the seeming infinity of the beetle world.

The tribe Aegialiini provides a case in point. It has two sub-species, according to a researcher that anoints the literature with the name Z. Stebnicka and his seminal 1977 work, *A Revision of the World Species of the Tribe: Aegialiini.*

This work deals with cold climate tribe members and has proved so definitive that it seems only to be amended from time to time, never replaced. Even so, the biology is still imperfectly known, even by experts. Aegialiini is not an important family in our yarn, and I have no intention of going there — there being the esoteric chambers of the total diversity known colloquially as dung beetles.

The focal point for almost every situation is the genus and species. The species holds in escrow a potential — if there is one — for a usable common name. It is the species that calls up its evolutionary history. It nails down a certain geographical location, writes its biological history, excludes its Darwinian relatives, and asks the masters of nomenclature to evaluate biochemical, behavioral and morphological aspects together with sexual reproduction of survivable offspring.

The International Code of Zoological Nomenclature adopted by the International Union of Biological Sciences is the modern bible for adding new species, new names and the mix of Greek and Latin. The guardians of Nature's exquisite mix are the miscellany called museums.

It is said that Aristotle led off with a biological naming scheme even before the time of Christ. He invented the name "coleopterist" to describe insect specialists, but the enduring system now used has its own biography. Catalogs appeared as early as 1876, one with 77,000 entries. A successor with supplements, about 1910-1940, listed 221,500 species. This appears to be the last real attempt to give beetledom a telephone directory style treatment for the world's beetles.

First, there is the matter of reproduction. Let's start with the wild onion as an example. It reproduces by seeds and also by underground bulbs, but since these systems might fail, the wild onion relies on its insurance mechanism. It has small bulblets at the base of the flower stalks. Often, flowers are missing, which means that the flower head simply has a main stalk with a cluster of small bulbs. The bulbs ripen and shower the soil with an onion squall, so of course the wild onion proliferates. This triple whammy for reproduction is hard to control, whatever the method.

Withal, most weeds are nevertheless reproduced from seeds and the plants almost always have seeds aplenty. Hundreds of thousands of seeds are more the rule than the exception. Fat hen, knotgrass, black bindweed, field pansy, persecaria, chickweed, speedwell, goosegrass, plantain, spurry, sandwort, scarlet pimpernel, toadflax and mayweed not only produce an abundance of seeds, they also provide nourishment for birds. These birds then pass seeds through their digestive tracts and, as Nature would have it, some few survive digestion and are unloaded, their shells humbled by digestive acids and ready to grow. Deposited on the soil, they increase and multiply, defying the wiles and laws of men, and mocking those of little vision who would legislate biology. Fortunately Nature has decreed that no weed should have a free rein. Like government they need checks and balances which they have in abundance if we have the wit to see.

Maybe this unwillingness to be governed helps weeds turn outlaw and take on aliases. Weeds are not only the world's greatest travelers, they are perhaps the world's most practiced con artists. They change their names as they march across county, state and national lines and pretend to be members of families to which they don't belong. Even the practice of giving weeds a somewhat Latinized moniker hasn't made the tribe turn entirely honest. The idea of genus and species for each plant and animal in a sort of coined Latin was born in the fertile mind of one Carolus Linnaeus. Historians tell us that his name was really Karl von Linne, but Latinizing his own name seemed a suitable precursor to naming each plant, insect and animal from a classical tongue. Carolus Linnaeus was born in Sweden in 1707.

In those days, people chose a Latin name as part of the Christian baptism ritual. "Linnaeus or Linne are the same to me," he once said. "One in Latin, the other Swedish." Since he bylined his writings "Linnaeus," naturalists came to know him by that name. He wrote some 180 books in his lifetime, which ended three years after the start of the American Revolution, or 1778. Archivists refer to either Genera Plantarum or Species Plantarum as his magnum opus. In those literary outpourings he mined everything in print starting from the works of Pliny the Elder and onward, including the massive three volume work of Tournefort with its name that can hardly be translated to English unless we accept a compromise and call it *Principles of Botany*. There were others too numerous to mention, most of which recited descriptions rather than giving names to plants.

Linnaeus was the first to define kingdoms, animal, vegetable, mineral. All plants belong to the kingdom called plantae. Dung beetles belong to the animalia kingdom. Each kingdom is divided into phyla. Phyla are then divided into classes, which are divided into orders, which, in turn, are made up of families. Families are divided into genera, or a genus, and last comes species: 7 divisions in all, sometimes more.

The *Onthophagus taurus* dung beetle can be classified as follows, Kingdom, Animalae, Phylum, Arthropoda, Class, Insecta, Order, Coleoptera, Family, Scarabaeidae, Genus, Onthophagus, Species, taurus.

In order to read this book, the reader is encouraged to stretch a bit, to shun not a few strange words that really define themselves and then relax with the flow of the narrative.

First there is the matter of Scarabaeidae, the main dung beetle discussed here. It finds application in identifying to three members of three sub-families of Scarabaeidae: Aphodiinae, Geotrupinae and Scarabaeinae. These sub-families are similar in that they live in the soil or in feces. Almost all of the dung beetle insects rely on consumption of excreta from vertebrates for nourishment and life propagation. Here we'll dwell on Scarabaeinae because of their penchant for coprophagy. The term defines itself for any boy who took a stick to stir a cow patty or perform necropsy on chicken dung or a rabbit's pellet. We will mention that a few beetles opt for carrion, organic matter in a forest or foods more often mentioned than defined.

The word "coprophagy" labels our quest. In all probability, dung beetles rely on the microorganisms in dung. This consideration elevates excreta to prime status not only for the coprophagous beetle, but for the soil and the crop. We can now cite the other feeding habit styles, necrophagy, mycophagy and saprophagy. All fall outside the purview of a story that, after all, starts with scarabs and remains open-ended. It may not be necessary for a farmer or rancher to memorize any or all of the above. It is enough to know that there is a structure to the world of beetles and that entities meant to ring the cash register can best be defined as genus and species. As for the sub-family Scarabaeinae, the latest references cite 4,500 species in 200 genera. When we talk of dwellers, ball-rollers or tunnelers, we are required to mention Scarabaeinae.

Not everyone agrees, but I like to write genus and species with the first word capitalized, the second word lower case, both italicized. Genus by itself takes a capital letter sans italics. That's the style we used at *Veterinary Medicine* magazine and, as I feel about most things classical, I consider it a sacrilege to change.

Some of my readers wonder aloud how it is that I love idioms and the vernacular while still insisting on the flavor of the old church tongue for certain purposes. I won't argue that the *Requiem Mass* by Wolfgang Amadeus Mozart is the most beautiful piece of music ever written (which it is) because we have to allow each his own. Some people prefer music that is to sound what a bran-stuffed Holstein's issue — backed to a barn door after a heavy clyster — is to art, and this is their own business. But Latinized names for each weed and beetle make it possible for competent persons in every language to communicate and understand. In this sense, Latin is a living language. The craft of the systematist requires it. Linnaeus saw that there were simply too many vernacular names in any case, each weed, each animal, each insect, each alias masquerading as the real thing — except there was no real thing. Geographical names meant little.

The English language is in a dilemma nowadays because of the drive to desex the sentences and paragraphs we use. Any editorial desk receives constant advice to use bastardized terms like chairperson and womanager instead of chairman and manager, for instance. This has never been a problem in Latin. It may be considered a dead language, but it has lively expressions. It is an inflected language, meaning that words change form to denote different genders. If the ending is —us, it means that the noun is the subject of the sentence. If it ends in —um, it means that the noun is the object. All nouns are either masculine, feminine, or neuter. *Ceanothus americanus* is a masculine name. *Cimicifuga americana* is feminine. And *Narthecium americanum* is neuter.

This much said, we must now face the reality that certain weeds have too many folk names, and sometimes different weeds and beetles, lay claim to the same common name. Two weeds claim to be the genuine tumbleweed, yet each has a distinct Latinized name. This accounted for an agonizing decision in writing these chapters whether or not to recite the scientific nomenclature each time a dung beetle is mentioned.

One thing is certain, dung beetles are sublimely indifferent to man's linguistic entanglements. They have their marching orders — a built-in computer that tells them when to sleep, when to wake up and when to procreate. Their coded DNAs have yet to be interpreted by science, biological propaganda to the contrary.

The absolute truth is that farmers generally do not recognize dung beetles the way they recognize weeds. Some speak of all dung beetles as tumblebugs, hardly pausing to distinguish between dwellers, rollers or tunnelers. If the difference was as evident to the unaided eye, there might be more common names, making reports on these unpaid manure dwellers and haulers more readable.

When Truman Fincher and his associates at College Station, Texas set out to field-colonize dung beetles across the southern tier of states in the 1970s, all of the species were considered exotic, and their collective action proved the feasiblity of controlling horn flies via the agency of specific dung beetle releases into the 1980s.

Our cast of characters starts with these, becomes enlarged as we range further afield, and becomes more definitive in the last chapter, which lists a few of the thousands of genera and species names as a token bow to the thousands that crossed my path as I followed the dung beetle trail. The roster reveals genus, species, date of release, country of origin and a few side comments.

Onthophagus gazella, 1972, South Africa to Australia: the king of African dung beetles. It has proved itself in the U.S. as far north as Oklahoma. One of the 15 released by Fincher's crew.

Euoniticellus intermedius, 1979, South Africa to Australia: it ought to have the common name "celly." Another of the 15 released by Fincher and associates.

Onthophagus bonasus, 1980, Pakistan: we see a good common name here, *bonus*; one of the famous 15 released under the USDA program of that era.

Onitis alexis, 1980, South Africa to Australia: how can we help but call this one "alexis"? It is a capable handler of cow manure if colonized in sufficient numbers.

Liatongus militaris, 1984, South Africa to Hawaii.

Onthophagus taurus, 1985, supposedly Europe to Florida: this horned fellow may have made it to the U.S. with imported cattle before it made it to Truman Fincher's roster for field colonization. It shows great promise for colonization beyond the near temperate zones of the south. Common name even now: taurus.

Onthophagus sagittarius, 1985, Sri Lanka to Hawaii: joined to the 15 that were field-colonized. "Homer" might be a good common name. Surely this one was around before The Iliad or The Odyssey.

Gromphus binodis, 1985, Argentina: one of the species that became part of the release program that threatened to resist the scope of dung beetle cleanup activity in the U.S.

Gromphus lacordairei, 1985, Argentina.

Onthophagus binodis, 1986, South Africa to Hawaii.

Onthophagus depressus, 1987, South Africa to Georgia.

Onthophagus nigriventris, 1987, Kenya to Hawaii.

Ontherus sulcator, 1987, Argentina.

Copris incertus, 1987, Mexico to Hawaii.

Sisyphus rubrus, 1987, South Africa to Australia.

Onitis vanderkelleni, 1987, Kenya to Hawaii.

Zinjanthropus boisei: the zin- prefix means "out of East Africa," where Louis Leakey found the specimen of Early Man. It is included because few Linnaeus-type names for dung beetles fix their origin by name. Most of these Scarabaeidae have been used to mark the various chapters of this book. Their diverse nuances were captured by the artist, Skeeter Leard.

Recall that short roster of weeds together with their Latinate names presented above! Now contrast the quality of communication possible when I mention the names of dung beetles packaged by kingdom. Only one of the above was identified and named by Linnaeus. It alone of this small cast of characters has a common name, one more or less assigned by laymen, namely "gazella."

It is not likely that very many common names will ever be supplied by farmers. If I had my way, I'd find names, perhaps "Oink Smith" for *Copris minutus,* a dung beetle that Truman Fincher described which prefers pig manure above all else, this based on a population explosion of the beetle at a swine farm. Tongue in cheek, could not *Onthophagus pennsylvanicus* simply be "penn" or "pencil" (pennsyl)? Or *Aphodius sallei* simply be "sally?" It won't happen, of course, because farmers and ranchers won't trot out the reading glass or microscope and the color manuals to identify more than the handful necessary for the dung removal chore.

The cast of characters that interests us is not limited to workers in our midst. Paleontologists have turned up dung beetles that pre-date the last Ice Age. The Smithsonian entry entitled *North American Representatives of the Tribe Aegialiini (Coleoptera, Scarabaeidae, Aphodiinae)* has chapter and verse on species both current and prehistorical. Although the title sounds like the product of a committee, the tract is valuable for the insight and detail that it offers. At the risk of smothering the rancher dung beetle aficionado in more Linnaeus scholarship, it nevertheless becomes necessary to note that the tribe Aegialiini has two sub-tribes: stebnicka aegialiina and eremazina. Aegialiinia (note the –a ending) is the only American sub-tribe. This means that six of the nine world genera are as American as apple pie. Here we have species adapted to a cold climate.

Linnaeus did not figure much in the naming of dung beetles. In fact, our special bug didn't really attract much academic attention before the 1930s. New species have been discovered and named ever since. The U.S. Entomology Laboratory in Washington, D.C. sees to it that each new scrap of information finds its way into the appropriate compendium. In other words, there are distribution maps and locality records that boggle the imagination. In the last case the range of each species is detailed.

The old saw that a picture is worth 1,000 words may be true, but it is also true that on-scene demonstration and instruction is worth 1,000 pictures. No one discussing dung beetles can become comprehensive without running into dozens of institutions and hundreds of investigators, a procedure that could easily consume a lifetime.

Suffice it to list here a few of the species alluded to above, with notes deferred to the Afterword part of this book. *Micraegalia pusilla,* for instance, has not been found except in the fossil record near Norwood, Minnesota. In really technical records, genus and species are written in the usual italic, followed by the name of the researcher who first identified and described a particular species. *Aegialiina rufescens* Horn thus names the worker who broke into print with his in-depth description of the anatomy, life tenure, habits and sexual expression of this newly discovered species. There's enough Latin in these names, some of which go back to 1775 or thereabouts, but most of them define the great era of dung beetle discovery from the 1920s to the present.

No cast of characters would be complete without the names of the Australian imported workhorses. The Onthophagus genus with species as follows: taurus, binodis, gazella and nigriventis were to be expected. Onitis answers the roll call with alexis, aygulus, pecuarius, caffer and viridulus. The Euoniticellus genus became a Down Under survivor group with intermedius, africanus, fulvus and pallipes finding their ecological niche required for survival. *Liatongus militaris* became one of the survivors listed by CSIRO. Three more beetles were illustrated in the CSIRO color-plated booklet, namely *Sisyphus spinipes, Geotrupes spiniger* and *Hister nomas*, a predator from the Histeridae family. There were other imports but these were the most successful.

Maps constructed by CSIRO showed where on the continent these beetles would survive, a total of 24. John Feehan, the sole distributor of dung beetles for Australia, offers a free identification service. All you have to do is scoop them up, float the bugs to the surface in a pail of water and mail them in a package that breathes, Feehan advises.

So what have we learned about our cast of characters? The careful reader will discern the mixes of Latinate word endings denoting sex and some few other factors, but without a narrative run on each length, color, description of jaws and at least a hint of the manure type preferred, we seem to say that our cast of characters makes a Cecil B. DeMille movie take on Lilliputian dimensions. Without achieving encyclopedia status, many of the identifying data will be answered in a general way as this story continues.

First, however, we might do well to expand the cast of characters by simply citing native dung beetles found in Kleberg County, Texas, chiefly on the King ranch:

Canthon imitator, ebenus, probus, vigilans, cyanellus, perplexus, viridis, puncticollis, lecontei, and *praticola;*

Phanaeus difformis and *triangularis ;*

Onthophagus hecate, pennsylvanicus, medorensis, landolti texensis, schaefferi, and *oklahomensis ;*

Aphodius lividus, haemorrhiodalis, rebeolus, and sallei;

Ataenius platensis;

Copris remotus;

Geotrupes opaca.

The roster varied somewhat, some different species and some missing, in Victoria County to the northeast and in Gillespie County to the north-west, and would have presented an awesome data base if developed for the over 2,000 counties in the U.S. (3,086 if we include Louisiana's parishes and Alaska's boroughs). Perhaps a glimpse of the other great dung beetle experimenter, Australia, will flesh out the picture.

The Aussie dung beetle program and its demise ran more or less paral-lel to the career of our own metaphorical and real dung beetle investigator, Truman Fincher. Shortly before funding ran into the trip-wire that sent the program over its own Niagara, CSIRO published a handsome full-color booklet entitled *Common Dung Beetles in Pastures of South-eastern Australia.* It is an awesome publication that covers its own cast of characters without the expensive frills that extension of its outline might have generated. It gives to the world a diagram useful to the average farmer and includes color photographs useful to the entomologist studying each taxa, and it adds to our necessarily brief list of the cast of characters hopefully of benefit to farmers and ranchers. One entry should illustrate the format of the CSIRO manual:

Onitis aygulus

Length: 20-25mm.

Colour: Pronotum dark brown, sometimes with green sheen; wing-covers light brown.

Special distinguishing feature: In males the hind femur has a double spur, while the female has no spur.

Distribution: The species occurs in the cooler, drier parts of South Africa.

Introductions to Australia: Releases were made between 1977 and 1982 in NSW, SA and WA, and the species is now established at some of the release sites in all three states.

Activity period of adults: From spring to autumn, ceasing temporarily during hot, dry summers.

Flight time: Beetles fly to fresh pads at dusk and dawn.

Biology: In moist, sandy soils this species buries its brood masses about 16cm deep. Each sausage-shaped brood mass contains up to eight eggs, with several brood masses per nest consisting of 500-700g of dung. Development time from egg to adult during summer is 2.5-3 months, during winter up to 10 months, and this is greatly extended during unfavourable periods. There is one, possibly two, generations a year. The species probably over-winters in the larval stage.

CSIRO goes on to give the data on colorfully pictured beetles, such as *Euoniticellus fulvus* with its yellow, tan and brown color blends and dozens more. The cast of characters is much like the Australian rabbit, see one, and you can be assured that a dozen or more are out of sight.

Of the 7,000+ species identified so far, only a scant few have been mentioned in these pages. The reason becomes self-evident when we recall the early experiences of our model entomologist, Truman Fincher. As a university student, he took on the project of measuring the fecundity of native beetles in the Tifton, Georgia environment. His trapping methods were hellishly ingenious.

He made *Phanaeus vindex* a household word among dung beetle enthusiasts. The same became true for *Phanaeus igneus, Dichotomius pinotis, Phanaeus carolinus, Copris minitus* and *Canthon pilularius* with a gaggle of species in tow.

In order to measure dung preference, it was necessary to trap the beetles. That process in itself revealed the character, stamina and secret lives of the beetles. As expected, the sexes were more or less evenly divided. There was, however, a profound difference between those captured on a swine or dairy farm and those in a laboratory setting. The vindex and igneus species preferred swine feces over all other feces. As Fincher's index took form, it became evident that situs governed capture.

These findings threw a new light on a profession generally focused on cattle dung for obvious commercial reasons. A data tug of war ensued as data on vindex and igneus filled the records and charts. The cast of characters identified itself by its flight pattern, its method of orientation and its preference for either daylight or moonglow. The collection process told all or almost all.

One can imagine the manhours spent emptying pit traps and marvel at the procedure. By the 1960s, the profession had invented a Rube Goldberg device called an autotechnicon. It held 12 stations, each holding a #10 coffee can. It rotated once an hour. Each unit had a paper cup and a funnel that admitted entry to the dung beetle, but not an exit, except for one devilishly clever species that found a way to escape.

On a swine farm it clicked off the hours. Each trap told its own story. It told when to hold and when to fold. It trailed activity the way a buzzard trails death. It explained to researchers when flight fell off and when activity closed down for the night. Only the Onthophagus and Histeridae species remain active after 9 pm. Most species open shop between 7 and 8 am, peak between 9 and 10 am, and close down sometime between 11 and 12 noon. After a siesta lasting until 7 pm, activity picks up again. The evening peak comes between 8 and 9 pm. Fincher carefully identified beetles according to minute details.

They are all individualists, these dung beetles, but they do have conductors, these members of Nature's silent orchestra. One such conductor is temperature, another is the bombardment of sub-atomic particles under the direction of light photons. For these reasons hourly temperatures were made a matter of record. Radiation per hour was taken from the Georgia Coastal Plains Experiment Station's meteorologists. Solar radiation was measured in Langleys or gram calories per square centimeter of irradiated surface. By the time Fincher finished his dissertation he knew more about dung beetles than a Catholic theologian knows about the feast days of the calendar.

We say that certain dung beetles are native, but we really don't always know. Perhaps some hitched their way to America with Ponce de Leon or migrated east after Hernando de Soto passed by out West. With a world-wide cast of characters of over 7,000, almost anything can happen and occasionally it does.

The rancher listens, at times fighting glazed-over eyes. What indeed can he do about rainfall, Langleys, temperature and the refinements of in-depth research? It suddenly comes to the cowman, now wide awake, that he must provide an environment in which dung beetles can live and feed. After the toxins are gone, the dung is provided via intelligent pasture management and ecologically sound husbandry. Rainmakers come and go and cloud cover abides. The hands of the gods stroke and they chastise.

Early on Truman Fincher found that any environment will have only a small inventory of beetles in terms of species. Often a species could be found in only three counties, or even in only one county. These thousands of dung beetles labeled by entomologists often frighten the rancher who fears that learning about the locals might be impossible, or worse, that he might exhibit ignorance and illiteracy, yet the cast of characters, like politics, is local. Like rainfall east and west of the isohyet, the numbers increase and decrease as choreographed by Nature and endorsed by man.

I close this chapter with a quote from Truman Fincher's thesis, *Ecological Studies of Dung Beetles (Scarabaeidae: Coprinae) Serving as Intermediate Hosts for Nematode Parasites of Swine in South Georgia (with reference to Beetles of the Genus Phanaeus).*

"Overall, the data demonstrate that dung beetles, in particular *Phaneus vindex* and *Phanaeus igneus,* are attracted more strongly to feces of swine than to feces of other animals tested, based on attraction index . . .

"New species records for beetles that can serve as intermediate hosts for *Physocephalus sexalatus* were: *Aphodius campestris, Ataenius simulator, Ataenius figurator, Ataenius platensis, Canthon vigilans, Canthon nigricornis, Copris fricator, Onthophagus concinnus, Onthophagus depressus, Ateuchus histeroides, Geotrupes splendidus, Phanaeus igneus nigrocyaneus, Phanaeus torrens niger,* and *Anomala flavipennis.*

"New species records for beetles that can serve as intermediate hosts for *Ascarops strongylina* were *Anomala flavipennis, Dyscinetus morator, Hybosorus illigeri,* and *Phyllophaga ephilida.*

"New records for *Gongylonema spp.* infections are *Phanaeus vindex, Phanaeus torrens niger, Pinotus carolinus, Canthon chalcites, Canthon pilularius,* and *Deltochilum gibbosum.*"

The ways of scholars are awesome, and the issue surely calls for respect. But let us return to cow dung and its connection to the bottom line, that equally awesome bottom line.

Liatongus militaris *is an import from South Africa via Hawaii, circa 1984. It is brownish orange with black head and forequarters, and has black striped wings.*

14
The Global Presence

The entomologist Truman Fincher tells me that some 7,000+ species of dung beetles world-wide constitute the going count, but even when the American program was terminated the number of unidentified beetles was still growing. On every continent except the South Pole ice cap there were survivors, some that could be dated back to the Paleocene Age, some perhaps marching and evolving through the Eocene, Oligocene and Miocene, meaning all of the ages that start with the geologic Mesozoic, Cenozoic, etc., meaning the age of mammals, reptiles, amphibians, fishes and invertebrates, in short, the business we call Planet Earth after 2-4 billion years, give or take.

According to Charles Darwin and his granddad and a Greek or two, life — the beetle included — invented itself and evolved and continues to evolve, as does any creature guided by a DNA and monitored by innumerable RNAs. Some few philosophers rely on the laws of chance and postulate that a principle exists that routinely defeats the laws of chance to create complex organisms.

It cannot be said that the dung beetle has gained much in complexity since the Pliocene. All that is known suggests that distribution of species, shaped for their climate and dung style, evolved according to subtle changes in the chromosome.

Truman Fincher and most of his dung-beetlizing associates believe that the most recent Ice Age pushed the dung beetles indigenous to the North American continent south into the tropics, for which reason there is a real paucity of extremely capable dung beetles in the U.S. even today. We may never know what happened to beetle fauna when worlds or fragments thereof collided.

There was such a collision, of course. About 55 million years ago, a fiery asteroid crashed into the shallow sea near what is now the Yucatan Peninsula. It was a chunk of rock from outer space about the size of Rhode Island. It had been traveling at perhaps 85,000 miles per hour, give or take, and lost its way for reasons only speculation can supply. It struck the land now called Mexico, creating a new geography and a new bay. The blow-back was so devastating that it cleared North America of trees and animal life-forms. It probably also wiped out dung beetles evolved from the Pliocene and left naked a land that complied with the ages of geology.

The crash terminated the age of dinosaurs, literally leveled most of the continent, extinguished species, annihilated woodlands and prepared the way for mountains to rise, for savannas to form and, not least, for mineral dusts to be distributed world-wide. One that asks for our attention is beryllium. It can't be found on land except at depths that invite paleontologists and their digging tools. Scientists date their finds by the beryllium layer which was uniformly distributed when the asteroid struck. As for dung beetles, we have to rely on the fossil record that paleontologists rarely supply, except for the sparse offerings of the Smithsonian. And these suggest a kingdom survival on a par with the cockroach.

A great deal of speculation attends all of these comments. Not so speculative is the fast-forward world that consumes so much paper and ink merely listing dung beetles by genus and species, and hinting at workers who identified the dung beetles and troubled themselves writing papers about them. Remember, the Jurassic Park theme was played out 130 million years before the tallest and biggest redwood tree, the General Sherman, got its start 2,600 years before the birth of Christ.

The global presence of dung beetles has made itself felt in terms of compliance with the Linnaeus scheme, common names eluding the bug the way common names elude bacteria and all microorganisms. The endless fascination attending the study and tracking of dung beetles has taken researchers to the far corners of Planet Earth. As noted earlier, the Texas releases that Fincher and associates took to the field included species from South Africa, Mexico, Pakistan, Sri Lanka, Argentina and Kenya with heavy emphasis on South Africa by way of Australia and Hawaii. It almost appears, to use Garrison Keillor's humor, that when God said, "Let there be dung beetles," he meant also that Africa should be their home, and that *Onthophagus gazella* should be the leader of the pack.

The prophet, perhaps even the messiah, of the global interest in dung beetles was George F. Bornemissza. Early on, he had already proposed attacking pasture pollution with dung beetles. He was born on February 5, 1924. He became senior research scientist at CSIRO, the Australian scientific agency, bringing to that post a European education that touched base in Germany, Austria and Hungary. There, farmers for many centuries had counted the beetle as one of Nature's blessings. After he came to Australia it came to him that Australian dung beetles were quite inefficient in clearing pastures of dung from domestic animals which had arrived along with the first boat-load of prisoners. This occurred shortly after the end of the American Revolution and Georgia could no longer be used as a prisoner dumping-ground. The arrival of European animals resulted in grassland pollution quite different from any experienced by that fragile terrain before. The effect on the environment need not be imagined. It can be experienced even today.

Bornemissza reasoned that if imports created the problem, then imports could correct it. He proceeded to introduce dung beetles from the Old World, beetles well adapted to recycling cattle dung, thereby restoring the balance of Nature. As stated previously, it was the business of bringing and breeding millions of animals that brought a new scourge to the Australian continent, face flies, buffalo flies, parasites too numerous to mention and economic losses that no one had pause to calculate.

Bornemissza estimated that three million dung deposits hit the turf every day nation-wide. Usually these deposits stayed on for months, even for years. The nuisance of flies was one thing, loss of valuable nutrients was another. This made dung beetle repopulation an international trade problem on a par with the new distribution of disease agents, now that POSH — Port Outbound, Starboard Home — has been replaced by air travel. No more maturation of a condition en route and burial at sea.

Hawaii invited in dung beetles of African origin early in century 20. Indeed, that island state became the conduit for African beetle transfer both to Australia and to the continental United States. Some of the names are jawbreakers, *Euoniticellus intermedius,* for instance. Seven thousand plus species of dung beetles world-wide would appear to be a good bank account, yet the global variables call into question the value of many artificial migrations. The U.S. itself has perhaps 150 species and most of them are inadequate for the job. Texas has about 50 species, many of them small-sized dwellers incapable of handling the freight that cattle-raising has put on their shoulders. The largest dung beetle on the global scale measures 3 inches in length by 2.1. The smaller beetles claiming dung-handling capacity are only a quarter inch in length, even smaller.

Matching fecundity and balance in one part of the world to another is no armchair job. Truman Fincher spent a great deal of time in Argentina trying to find even one beetle that fit the conditions of the Southeast. Answering the question once it has been asked often leads to tales of woe. The global scene has a diversity not easily comprehended. There are many predators of dung-breeding flies: Sphaeridium, for instance. Larvae of a dung beetle named *Sphaeridium scarabaeoides* (family Hydrophilidae) by the Linnaeus method have been reported preying on dung-breeding fly larvae, including the horn fly. Research adds to the above the intelligence that two exotics, *Philonthus flavocinctus* from Southeast Asia and *Philonthus minimus* from Africa, Asia, Hawaii and Australia were scheduled for Texas release in 1987. Unfortunately, the program closed down before definitive results were achieved. California had a 1968 release of seven exotic predators for horn-fly and face-fly control. They were *Atholus coelestis, Hister caffer, Hister chinensis, Hister nomas, Hister scissifrons, Peranus maindroni,* and *Santalus parallelus* (mostly non-Scarabaeidae). The trail came to an end after that and the survivors, if any, hardly made a dent.

Man proposes, but Nature disposes and those proposals leave tracks as wide as a forest fire. Earlier I promised not to burden this narrative with paragraphs peppered with the italic of genus and species, no common names being available. Still, a dozen titles of papers to suggest the global scope of the inquiry might help make a point.

The Nesting Behavior of Dung Beetles, Instituto de Ecología, Mexico City.

The Dung Beetle Man, Sydney, Australia (video with George Bornemissza).

Dung Beetle Ecology, Princeton University Press.

North American Representatives of the Tribe Aegialiini Scarabaeidae, Smithsonian, Washington, D.C.

The Scarab: A Reflection of Ancient Egypt, Israel Museum, Jerusalem.

Revision of Species of the Tribe Psammodiini from the Australian Region (Coleoptera, Scarabaeidae, Aphodiinae), Slovenské národné múzeum, Bratislava, Slovakia.

Dung Beetles and Chafers (Coleoptera: Scarabaeoidea), Royal Entomological Society, London.

Workshop on the Effects of Parasiticides on Dung Beetles, CSIRO, Brisbane, Australia.

On the Species of Canthon and Phanaeus of the United States with Notes on Other Genera, American Entomological Society.

Studies of Decomposing Cattle Dung and its Associated Fauna, Dissertation, University of Oxford.

The Sacred Beetle and Others: Excerpts from the Writings of J. Henri Fabre, Hazeltree Press, Charlottetown, Prince Edward Island, Canada.

The Bio-ecology and Control of Coproica vegans and Coproica hirtula (diptera: Sphaeroceridae) in Cattle Feedlots, Dissertation, University of Orange Free State, South Africa.

The coffee was old and cold when a former war-time buddy dropped in as I concluded this chapter. I handed him the manuscript. He read with increased discomfort as he came to the final page or two.

"Do you really think that farmers want to know that much about dung beetles?" he asked.

"Probably not," I said. "At least not the Linnaeus names." Indeed, tumblebug and scarab and perhaps gazella and taurus ought to do. I required my visitor to listen to my summary before winding it down. The American experience I said, could and should determine the scope of the future and the present.

Hard numbers literally cudgel the bookkeeper into submission over the add-on value of nitrogen. By assembling the studies of a platoon of experts, we can extrapolate these figures. For every 100 head of cattle, there is a potential loss of 851 pounds of nitrogen and 238 pounds of phosphorus. These numbers quantify the nutrient loss via desiccation and evaporation into the ambient air, with leaching and weathering forever squandering profits. This equals 24-34% of the pasture dung. Interment incorporates these resources into the plant-root zone where microbes take over. The resultant pasture improvement, red-meat production, and hundred-weights of milk extend the calculation until it reaches the profit line. A rancher may or may not compute the values attached to cleaner water, improved air and bucolic scenery, but all are values that cost or reward somewhere in the business and economic spectrum.

The paucity of dung beetle help is exacerbated by the nitrogen deficiency in most warm-climate acres. Dung left on a pasture to be desiccated by the sun has 0.4% of dry matter to contribute to the soil. This computes to an 80% loss of nitrogen. Yet that same dung taken underground enables the nitrogen to be absorbed by the soil colloid. State-of-the-art pasture management calls for about 400-425 nitrogen units per acre, but cattle feces can supply only 360-380 units per acre. The loss here is a mere 15%, not the 80% expressed earlier when dung remains unburied.

Fincher and a number of his associates worked out solid numbers with an actual experiment, not a speculative guessing game. They had conventional pastures. They measured. They colonized. They computed, and they expressed their findings in hectares. Official science may have gone metric, but American ranchers are barely conversant with the French Revolution measuring system. In any case, the results squared well with Australian findings.

In an age of inflation, it becomes useless to quote figures that become obsolete before the ink is dry on the spread sheet. Those who have lived to a ripe old age can remember gasoline at 14 cents a gallon and the $5,000 legal fee to Clarence Darrow for handling one of the greatest civil rights cases of the 1920s. Here then are some generalities that any farmer can flesh out with a hand-held computer.

A bottom line number for each animal becomes astronomical millions when extended to the national bovine herd. Total losses for the dairy industry due to problems easily remedied by dung beetles on the premises also tabulate into the millions. Internal parasites take their toll. For sheep, goats and horses, medication costs reached above the $100 million mark annually in the 1980s. Small pastures without dung beetles have 14.7 times more nematode larvae than plots heavily populated with a miscellany of dung beetles.

One has to marvel at the ingenuity of investigators when they set out to nail down a conclusion. Imagine screening off two acres so that dung beetles cannot migrate, then recording the infestation of calves, all with that markedly different environment. Of course, the calves with a dung beetle assist provided hard numbers proving the efficacy of the beetle in bovine health maintenance. A 31% decrease in worm population is nothing to be sneezed at, especially when aligned with the 9.9% increase in the national worm population.

Sadly, only 9 of the 40 species in the Southeast are capable of burying the overload of manure in their territory. Only 3 of the 9 became plentiful during the golden era of repopulation. The remaining 6 species were and remain limited in their distribution. All of the above invite both conceptualization and completion, even computerization, of that greatest of inventions.

As early as 1974, it was computed that the value of dung beetles to the livestock industry was in excess of $45 billion. A decade later, it was computed that losses inherent in slaughter-weight records came to at least $15.25 per animal. Such numbers make lines on a graph dance without rhythm, yet they tell the rancher that he or she must look to canceled costs and maximum gain for a bottom line and not to public policy.

We are not allowed to forget morbidity losses, more millions down the tube. Excess labor, extraordinary veterinary costs and special paraphernalia — add more millions, and then factor it back to the number of cattle and acres. There are no economical means for destroying parasites and their larvae unless that free labor force is summoned.

This is the picture not only in the United States but world wide.

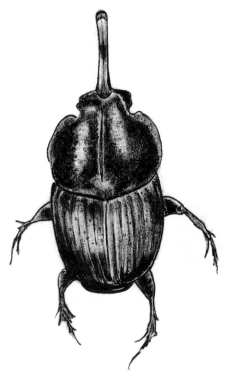

Onthophagus nigriventris *is a native of Kenya and other parts of Africa. It was imported via Hawaii in 1987. Its head and shell are charcoal brown with a few small speckles. The wing covers are sorrel and smooth, but only slightly shiny.*

15
Soils & Pastures

Down Nowland Road a short piece from where I live, a home-lettered sign announces to passing motorists: Rich Topsoil For Sale. "The man ought to be arrested ... buggy-whipped," one of my neighbors says. "That stuff is pure clay, about an inch removed from being cement."

The pitchman at the site rhapsodizes, "We've pulverized it for you."

Yes, I said sotto voce, you've reduced clay to a powder. When water brings the particles together you won't be able to break it with a sledge-hammer. A dung beetle couldn't penetrate the stuff except in soggy weather.

That seems strange because soil starts as fine rock powder. Biotic life donates its bodies to the topsoil conditioning process and fresh cow dung, that prince of nutrients, delivers the phosphorus and nitrogen that plants crave. Grass requires trace minerals which are often present but unavailable. It is the function of microorganisms to act on earth minerals making them water-soluble. Chemists pretend to perform this function and they do so with mixed results. All of the nutrients — major, minor, and trace — concern us, as does tilth, carbon, humus and organic matter, but the mystery of nitrogen overpowers both reason and the senses. The dung beetle's role in nitrogen production may be one of the most overlooked factors in economic pasture management.

Soil type governs the success of imported species. However other variables also figure in, one of which baffles the investigator. Feces are released throughout the entire day, but many beetles have flight patterns that affect the arrival at the payload. We tend to discover absolutes and then further inquiry knocks them into a cocked hat. So cattle dung would seem to be high on the preference list. Not so, says Truman Fincher. In the Georgia Southeast he found that many beetles preferred swine excreta even when cattle dung was available. Such a finding does not offend the reality that on the high plains and temperate prairies of the northern hemisphere, beetles definitely prefer cattle dung. Excreta from man, sheep, horse, donkey, camel and goat follow in that order, but not necessarily in that order of importance. Man's assault on the environment — especially on forests and grasslands — with cities, roads, dams, airports and spray planes, has created an ocean change in species and numbers.

Dung beetles, often considered homesteaders of the forest, are in fact most numerous in open pastures. If the grass is luxuriant, rollers need not apply. Scientists with the vision of ecologists, or with a clear view of evolution, think in terms of earth upheavals. The separation of North and South America must have arrived 105 million years ago, a bit after the humanoid that Louis and Mary Leakey discovered. Earth's angry moves touched Australia 65 million years ago, approximately at the end of the Cretaceous period. Tying dung beetle development to the movement of the earth's crust may be of no more than academic interest to a cowman but it seems to captivate the men and women who often spend their lives on the trail of families, genera and species of dung beetles. Yes, indeed, South America separated from Africa 105 million years ago and it is now a given in high-school classes that the complete isolation of Australia occurred by the end of the Cretaceous period. The point here is that climate changes with the movement of the earth's crust. That gifted split from Africa seems a country mile removed from tectonic lore, but is it?

The earliest of the dung beetle seekers wanted to start the way that James Michener started his novels on Hawaii, South Africa, Afghanistan and Alaska — "and the earth cooled." The nutritional ecologies are quite different when local residents are compared to the immigrants under college auspices. If someone took the trouble to match and measure the DNAs of strangers, the relationships could easily duplicate the genome work that now somehow relates a Norseman to an Aztec warrior.

Withal, the beetles seem to have sorted themselves out. The smaller species gravitate to colder climates. We know too little about their sexual transfer of genetic materials, but it is not out of order to think that cold often determines sex, as it does for turtles that use cold to create male progeny, exactly the opposite of crocodiles that use cold to create female progeny.

Larger beetles dominate wherever they are. Could it be otherwise when a contemporary elephant unloads enough dung to provide a day's work for waiting gazella beetles? Some have broad climate tolerances, yet lack the competitive spirit so beneficial when more than two species conspire to make short work of a cow pat.

The entomologist can identify the variants and the species, cranking out approved Latinate with the skill of a circus barker. Here, it is enough to know that the immigrants must match their culture to the new one when they are taken from their ancestral habitat. This was a challenge of awesome dimensions when the New Orleans-approved dung beetle project got under way.

Some ecologists know the Jurassic Age as the Scarabaediae Age. Some of the feeding habits have been connected to tropical forests where dung from small animals asked for attention. Indeed, some species still prefer small-animal dung. Marsupials in North America provided gourmet dining for the style of beetle we now call scarab. No doubt diversity took over as climate shifted to comply with global cooling, or perhaps global warming, such as occurred around 6,500 B.C. when ocean ice melted and raised the sea with a resultant rupture in the Bosporus land bridge which created the Black Sea. We can rest with these few notes and fast-forward to immigration in or around the A.D. era, often referred to now as C.E. (Common Era) by many historians, geologists, paleontologists, etc.

Our interest, however, leans toward those dung beetle immigrants that prefer the grazing space of cattle. Of these we sing. We accent the scientific speculation that dung beetles date back to the Mesozoic Age with fair differentiation during the Cenozoic. The die for the family seems to have been cast during the Oligocene, Miocene, Pliocene and Pleistocene epochs.

Carbon sinks, global warming and dung beetle habitats notwithstanding, the main reason for trying to understand the dung beetle's divine commission has to do with soil management, crop production and animal husbandry. Walt Davis, the stalwart cattleman and dung beetle host on his Oklahoma ranch, put it this way, "Loss of production due to weed infestation is of major concern to farmers and livestock grazers."

Walt goes on to assert that this concern has created a multi-billion dollar world-wide market for chemical and mechanical weed control products. The American penchant for throwing money at a problem does not seem to have worked. Its sole aim has been to kill the weed, burn it down, annihilate that "plant out of place." Unfortunately, this futile mercantile exercise does not deal with cures and leaves in place the bad management practices that offended the soil in the first place. The astute Oklahoman holds that weeds are a symptom, but not the cause of the problem. He lectures, he advises fellow farmers and he sells new paradigms, one of which is the role of this natural construction engineer, the dung beetle.

In fact, dung beetles and weeds play a most beneficial role in balancing the soil, plant and animal complexes that are called pastures. When dung beetles are missing or inactive and weeds proliferate to a point of crowding out forage species, then the mixed signals of fecundity and balance falter. Animal production pays the price.

The nay-sayer demurs. What about poisonous plants in the pasture? Do dung beetles scatter the seed and undo their office? The on-soil observer says, "No!" and "No!" again. Healthy pastures do not give poisonous plants very much permission for life. Health declares the kingdom for the season and healthy grass depends on soil with balanced cations and the inventory of attributes that follow dung beetle burial of manure the way a robin finds worms — with diligence.

Soils invite certain plants much the way that soils combined with climate order up the arrival of dung beetles equal to the task, all in the fullness of time. Yes, there are anomalies that offend our razor-sharp common sense. Nature abhors bare soil, and yet the scarab tumblebug functions best when it can roll its larder like a golf ball and therefore shuns lush grass. Then Nature plugs in the tunneler which sees to aeration of the soil, installation of nutrients in the rootbed and holds in escrow the nitrogen that, when liberated, changes the atmosphere on a scale often assigned to automobiles.

When the operative dung beetle changes the soil, it also changes crop conditions. The climax crop of grass is possible only when the soil is in climax condition. "When growing conditions are improved," Walt Davis tells all who will listen, "the succession of plants most valuable for grazing will advance. When those conditions deteriorate, succession will regress. Those plants most valuable for grazing are high successional plants that share common characteristics."

Those plants are wide-leaved, deep-rooted, long-lived and capable of converting sunlight, water and minerals into plant energy. Those photons of light match the hard minerals of the planet's makeup, sub-atomic particles and traces balancing each other.

Bounty invites abuse. Lush grass production tempts farmers and ranchers to overgraze. Rape of the pasture, even with dung beetle repair work going on, tends to lower the vigor of successional plants, generally cancelling out plant numbers and driving them backwards. Accordingly, the capacity of the soil to take in and hold moisture impedes the recycling of minerals in the soil-plant-animal complex.

Weeds are low-successional plants. They survive under starvation conditions. In short, they stake their claim on poverty acres and poverty soils. This phylum of plants are not to be dismissed. Their function is to assist the soil-building process. They contribute their root and plant systems, provide food for microorganisms and otherwise mix their grass with the fragmented manure that dung creatures provide.

Dung beetles improve fescue and Bermuda grass and also support the water cycle. Plants take trace minerals from the air via the stomata and install them into plant tissue, finally into the meat protein that grazed on Nature's greatest benediction — grass! Thus, the beetle joins forces with soils to reduce compaction and soil erosion.

Wes Jackson's *Altars of Unhewn Stone* reports that a good pasture will have perhaps 50 species of forage. It is never a monoculture. Each of these species has a different need and special ways of satisfying that need. Thus, we have warm- and cool-season plants just as we have warm- and cool-climate dung beetles. Withal, it takes diversity to utilize all the resources, air, water, minerals, dung.

There is an old Vermont farmer's saying, "If you find weeds invading your grass sward, fertilize! Weeds are a lot like human beings and civilization. Make them too prosperous and they perish." When the tunnelers go to work, the grass on top is saved and the weeds perish.

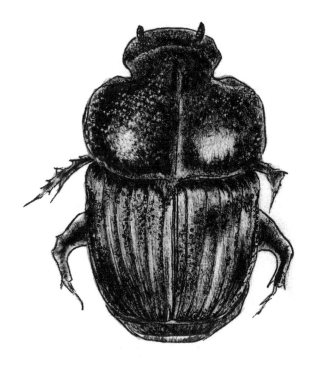

Onthophagus saggittarius *was imported in 1985 from Sri Lanka via Hawaii, and became one of the 15 colonizations that Truman Fincher and his associates accounted for. It is rust brown with a heavy black speckling texture on its head and shell, but lighter on wing covers.*

16
Five-Star Dining

On October 30, 2006, the Associated Press carried a report that should have delighted coleopterists, but didn't. The Miami, Florida zoo reported with pardonable pride the display of mock animal excreta. The headline was an arresting streamer: Miami Zoo Presents Poop Exhibit. Poop is not a word I have used herein, other words, both slang and official, being available, meadow muffins, horse/buffalo biscuits, bovine scat, bird guano, feces, solid waste, cow pat, cow platter, cow discs, even goop are a few of the names one hears for excreta.

To the above report, the Miami Metro Zoo added the term "educational." According to the press release, plain common scat became an exhibit item in a 5,000 square foot space with models of the real thing. The display announced "The Scoop on Poop," poop being understood by children the way that "tinkle" somehow sanitizes urine. Complete with pictures, there were realistic examples (sans odor) of football-sized elephant droppings, pellets that looked like kidney beans or garden peas, black fur-coated bear excreta and other samples that a full inventory would account for.

What an opportunity to exhibit the species of dung beetles that favored each style of poop as five-star dining! As a bow to the missing molecules called natural odors, a few flowers that emitted a dung-like odor to lure pollinating insects rounded out this educational exhibit, as did the ever-delightful scarab rolling a globe-like dung package up a synthetic hill. Admittedly, there is a lot that can be learned from a realistic depiction of a hippo marking its territory with precisely placed droppings. No doubt, children playing No. 2 I.Q. is also educational, but none of these educational tools match the daily dining rituals of dung beetles, whether taken as families, genera, or species.

When beetles slurp their dinner at a freshly delivered dung platter, they follow an unschooled directive ordered by their DNA. Without a day of instruction and with absolutely no credentials, they join the fray at the fount of nutrition and go to work. This is not to say that dung beetles are automatons. They have their preferences. Some turn up their anatomically correct snouts at pellets others crave.

In the eastern part of North America, fully 40% of the species described by entomologists are strictly deer dung enthusiasts. About 20% are micro-habitat specialists. Pasture occupants in that area are introduced species for the most part. There are dung beetles that prefer the droppings of marsupials, rabbits and every form of wild life under God's sun and others that would rather starve than reach for five-star dung created by swine. If there is a large animal in the neighborhood, you can make book that, given permission, dung beetles will be on its trail.

"Permission" is a code word nowadays, since many of our grasslands were sprayed with atomic waste after the Nevada tests in the late 1940s and early 1950s. In 1949, the United States established Poison Control Centers. Actually this was an announcement that henceforth agriculture would be going high-tech. Crops and pastures would be treated with salt fertilizers and the resultant carnage on plants and soil would be remedied with rescue chemistry. *The Farm Chemical Handbook* started out to be rather small and unpretentious, but before long it became an inch thick touting chemicals of organic synthesis — herbicides, pesticides, rodenticides, hormone sprays — an alchemy that could send a grown man into convulsions. The effect on the native dung beetle populations can be imagined. As the ranges filled up with cattle, the few dung beetles on the job were understaffed and hardly evolved enough to handle the ever-increasing workload. Importation of "king of the mountain" dung beetles such as gazella was a long way in the future and the fast-forward button was not yet working.

Some beetles developed a morphological adaptation to and propensity for exploitation of large animal feces. It is an article of faith and science that "that which came first" does not apply, since development of the best coincided with development of the five-star diner.

Herbivores graze in open habitats. They consume a prodigious amount of bulk and demand copious amounts of water as well as air. Always, the signal word is "excrement," timely delivered excrement, at least 8 or 9 times a day. Dependency means the specialization of which we speak.

There was and is an expectation of freshness cafeteria-style. The five-star meal loses its appeal faster than a Porterhouse steak in an over-refrigerated restaurant. Competition is intense, not only from competitors, but from desiccation in the ambient air. Competition is self-evident and fierce among dung beetles, thus the business of hydration if the meal is to be consumed at a later time.

Once the food has been located, attempts at underground burial begin, this to cancel out the ravages of heat and wind, both of which steal away life-preserving moisture. Down under is the refrigerator and the micro-climate, both of which thwart competitors, but predators are everywhere. The rapid removal of dung for later consumption lessens the risk for beetles becoming fast-food themselves for a quite different type of predator. Tunnelers and rollers each have their choreographed habits, as we have seen.

Real five-star dining often awaits a rainy season. A lot of dung beetle meals depend on the spacing of large animals. Rollers spend more time above ground than tunnelers. Consequently, their exposure to predators is greatly increased, although they are not entirely defenseless when their predators come to scoop them up. They are equipped with particular glands which emit repellents to their enemies.

Competitors, specialists, thieves, opportunists, military strategists, sometimes used to praise, sometimes to disparage, these are a few of the epithets, for better or worse, that one finds among the thousands of species catalogued by studious entomologists. Each can have and deserves its own biography, a proposition that schoolmen seem to attack with their customary alacrity. Farmers see this as an aside less important than the speed at which fresh dung disappears, permitting grass to shake off its rank growth and set about turning sunshine into raw material for more dung beetle food gain.

Reproduction seems to be species-specific, fecundity being reserved for the well-fed laboratory conditions that enable more than one generation per year. Our old friend the gazella is the five-star champion in fecundity, 180-200 eggs under lab conditions, according to Bornemissza. Most species deliver about 20 eggs per annum.

John Feehan, the sole distributor of dung beetles in Australia, knows five-star dining when he sees it. Although he specializes in beetles that like to dine on cattle dung he bows to neccesity when necessary. Urbanites like dogs and dogs are often taken to the nearest park to do their business. Feehan has field-colonized beetles in such a city park and the experiment proved successful.

For our purpose here, it is enough to recite what to some may seem trivial or obvious, that a native Australian species finds its five-star dining in the pellets of kangaroos. These pellets are approached from below and are not packed into tunnels. Down Under scientists tell us that this species exhibits a "nuptial feast." *Onthophagus taurus* also dines on kangaroo fecal pellets. It buries them intact and employs each as a brood mass. Another native dung beetle specializes in fecal pellets of the three-toed sloth.

I have mentioned species of Onthophagus that purloin the balls of dung rolled by other Scarabaeidae. The parasite thus finds revenue in kind, independent of great expenditures of energy. Some species favor excrement a bit more desiccated than the fare that some competitors prefer. Others construct brood balls within the confines of a single animal evacuation, the nest itself being limited by the mass of the dropping.

In other words, among those 7,000+ species identified so far, there are many that deviate from the norm expected of a bona fide tunneler. The quest for that five-star meal is a constant. There are species that have given up subterranean activities in favor of "just right" ecological conditions, not so much as panhandlers or kleptobugs, but deriving from the same force that dictates diet for almost every living form. They turn to drier dung at drier periods of the year. They change their nesting activities, not as tunnelers from which they evolve, but as opportunistic dwellers of sorts that attack their five-star meal from underground, especially cattle manure.

Wet manure slurpers do not like to compete with dry meal bugs. No cause is ever complete. Cause bestows cause so that we have proximate causes, causes once, trice, even hundreds of causes removed, until we arrive at that much-detailed First Cause, either the magic of Darwinism or the calculated cause of the mathematician and philosopher.

Researchers often lose themselves in that devil called detail and forget the reason for their dung beetle careers, namely the use of gained knowledge to assist the more practical business of meat protein production, refinement of that postulated "greatest invention," accounting, and finally, a bottom line that sustains ecologies, people and economy. In any case, a large food source is the handmaiden of the sex drive and the manufacture and survival of progeny, ultimately the open sesame to fecund population expansion and removal of those tons of manure from a grass sward.

Onthophagus gazella has been and continues to be the workhorse of the family, if not of the kingdom, presiding over Scarabaeidae. Its transfer from Africa to Hawaii to the mainland and its performance based on fresh cattle dung both tantalizes and frustrates. There must be dozens, perhaps hundreds of those gazella types somewhere on Planet Earth, beetles that so savor a cow dung meal that they rate attention as a removal mechanism on a par with Fletcher Sims's mechanical scarab machine. The data base supporting this transplant says that gazellas can bury a ton of manure per acre per day acting as teams. *Onthophagus gazella* remove 90% of the surface manure under conditions of heavy, rotational grazing and state-of-the-art pasture maintenance. Such a performance enables a scheduled return of the herd and rest and regrowth of pasture grasses and forbs. If gazella is exotic, then so is manure from healthy cattle. It entices the way a five-star restaurant entices the gourmet.

Such observations tend to cancel out the fact that cattle are not alone as meat protein producers. There are sheep, goats, horses in several cultures, ducks and other poultry, and not least, swine. Rainfall may monitor beetle activity, and lack of it grounds the nocturnal and diurnal flyers, but swine manure is often the gourmet meal of the lot. Truman Fincher was a young student when he noted that, at least in south Georgia, climate, humidity, and evolution conspired to make swine manure first choice fare for the dung beetle larder. Earlier researchers placed in the record the intelligence that for most of the northern hemisphere dung from cattle, human beings, sheep, horses, camels, goats and donkeys presided. That was in

1966. During his early researches, Fincher found that so-called progress — roads, urban sprawl, factory-in-the-field farming, herbicides, pesticides — all became responsible for a turnover of species available for service to animals, both wild and domestic.

Fincher used swine manure to evaluate dung preferences, and, for the Tifton, Georgia area, porcine excreta won hands down. Fincher wasn't interested in trivial pursuit. He wanted to know how dung beetle service and the beetle's dining preferences translated into biologically correct animal husbandry. His quest took him to the field, to off-shore islands undisturbed by an agriculture that had become conventional overnight, much like a fraternity initiation tradition that "goes into effect Monday morning." He concluded that if there is dung, there is a beetle to match — unless man disturbs the environment.

The discovery of a high preference index for swine feces was unexpected, yet reasonable. Historians tell us that Hernando de Soto explored with his commissary on the hoof. Having crossed the Mississippi, he tripped across the now southern states from 1539 to 1542. Along the way, some of his pigs managed to gain their freedom, disappearing into the wilderness, where they increased and multiplied. They survived in woods and rainbelt swamps so successfully that their feral descendants represent the only Spanish breeds still extant.

Most important, this infusion of a strange new manure into the environment caused Nature to order up a dung beetle that became quite partial to swine dung. Pigs trumped the manure output of bovines simply because of their presence several centuries before cattle arrived in force.

It may be that the presence of wild pigs is the most unrecognized trend in the U.S. today. Just as pigs arrived several centuries before cattle, so too did the beetles that came to enjoy this form of five-star dining. Wild swine may be a non-native species, but then just how long does it take to achieve "native" status?

Hogs cannot sweat but they can and do eat. Their dung is almost as special as human feces and their fecundity is equally impressive. In 1982, 18 states had wild hogs and presumably the appropriate dung beetles. By 1999, 9 more states became hosts to this porcine rotovator. By 2004, wild hogs had become established in 28 states, mostly in politically red states, as one wag put it, without too much scientific backing. As these lines are being set down, 3 more states have become wild hog heaven, Kansas, Oklahoma and Nebraska. Perhaps South Dakota will join the crew, this according to non-credentialed folks who make observations. All states with colonized wild hogs report exponential growth in hog population, even while the cattle herd is also increasing, all this according to *Wild Pigs in the United States*.

The Trox genus prefers dog dung, with fox feces a close second. Raccoon feces may have missed five-star-dining status by coming in third and opossum fourth. Trox does not seem to be interested in any other animal manure. Trox is from a Greek word meaning "to gnaw on dried carcasses and carrion." Their repast takes place after other coleoptera have had their fill. Spot situations in tropical areas have served up the intelligence that, in addition to dung, several species live on carrion.

The matter of dung preference became a target study long before Fincher made his reputation as an internationally recognized figure in dung beetle research. The starting point was recognition that wild animals populated virgin America long before man introduced domestic animals. The human species was also present, for which reason it was discovered that human excreta attracted dung beetles at least as much as that of wild fauna. It was this reality that governed procedures for the matter of measuring the beetle's feeding preferences. Habitat figured in, and the average number of beetles captured suggested an index.

Average number of beetles captured per day in feces-baited pit traps at three different locations. (Georgia Coastal Plain Experiment Staion, Tifton, Georgia, 1967.)

The traps were baited with swine feces and dung from opposum, fox, sheep, man, rats, raccoon, even chickens. The records were so meticulous they would make any bookkeeper wretched with envy. Tastes for five-star fare seemed to depend on species. It was always a case of location when the time came to count. At the end, it became evident that evolution had bestowed a balancing mechanism for open-air dung, and its name was dung beetle.

Attraction index became added to the lexicon of dining research. Not only food preference, but the full range of life pursuits and habits emerged from the student's study, food preference being only one element. The preference of swine dung over cattle droppings must have surprised the Tifton entomologists monitoring Fincher's work.

There is a huge amount of food for dung beetles in the United States. Why then, of the 197 species in North America, do 130 reside in Mexico and only 87 in the U.S.? Furthermore, of the 87 only 40 are found in the southeastern states where most livestock grazes year-round. Even further, of those 40 only 9 are considered capable of dealing with the ever-expand-

Beetle species	Average captured per day	% of total
Onthophagus (various)	238.4	66.38
Phanaeus vindex	84.86	23.64
Hister (various)	25.03	6.97
Phanaeus igneus	4.41	1.22
Pinotus carolinus	2.68	0.75
Geotrupes (various)	1.72	0.48
Canthon nigricornus	0.67	0.19
Trox (various)	0.64	0.18
Copris minutus	0.58	0.16
Canthon pilularius	0.10	0.03
	358.93	100%

ing dung production from ever-expanding cattle herds. And, only 3 of the 9 live on-site where most of the work remains undone. The native 6 species are subjects for school papers and Ph.D. dissertations, but they're not the troopers that the greatest invention of honest bookkeeping requires, nor do they provide the bottom-line profits that the stockman seeks. The availabilty of five-star dining and the paucity of diners have puzzled students of the tribe. This much stated, it becomes obvious what program consumption must follow. The key is favorable rainfall. This is what Truman Fincher and the program his associates championed sought to accomplish and what the klepto-political people sought to close down.

We tend to rhapsodize about the prowess of dung beetles in removing bovine dung, but we overlook the five-star dining potential of a layer of undigested material. John Feehan reports that he has counted as many as 1,000 dung beetles in one cow deposit. Interment, he relates, often takes less than several hours. The usual sequence of brood balls and deposit disappears on a Texas ranch in 24 hours if the weather smiles and bestows enough moisture to make the meal a working possibility. In a soft spring the pile does a Houdini disappearing act, only a shell of its former mass remaining, leaving any fresh excrement to the mercy of the sun and dessication which often occurs within 48 hours, the same time it takes for potato salad to spoil at room temperature. It isn't that the dung "spoils," it's just that the gazella doesn't like roasted food. That is why a horse delivers an unsatisfactory meal to a gazella.

Except for cultural dictates, there is no real reason that people who rely on meat protein can't consume horse, dog, cat, small game, even rodents. Why not beetles, beetle larvae and in many cases beetles whole? Certainly we already consume insects since it is impossible to keep them out of the food supply.

There are enough beetle recipes to fill a beetle cookbook. Gourmet beetles are harvested from logs and usually do not involve scarabs. There are exceptions. Some French and Japanese chefs offer their customers what they call roasted or boiled grubs, actually the larvae of melolonthine scarabs and cerambycids (longhorn beetles). In the main, five-star dining on beetles meanders into the many families — with or without common names — mentioned in an earlier chapter. (There is even a *Foods Index Newsletter* published by the Entomology Department at Montana State University.) We turn up our noses at scarabs for food or medicine, but legends abide.

The 15 species that Fincher and his associates studied leave unanswered a question not asked so far. What will be the effect of genetically modified grain on cattle dung as this unnatural feed asserts itself, distorts the digestion of ruminants and finally tracks its way into the fare of dung beetle diners? Do we have a clue from an incident reported by Fincher to a journalist?

It seems that a colleague of his was staying in a hotel in a provincial Indian town. He noticed that each morning men were disappearing into a lakeshore thicket. A local man told him that the area served as a toilet facility for indigents, even for the working population. The colleague concluded that the area would be a perfectly horrible example of filth, flies and diseases waiting to happen. As a scientist and investigator, he decided he'd better take a look. He found little human excreta. What little he did find was being diligently buried by dung beetles.

There is a rare condition, now a matter of record in India, called "scarabesis." It is caused by beetles of the genus Onthophagus always on the hunt for a favorite dung, human feces. These beetles become trapped in the rectums of human victims, usually children who fail to clean themselves after defecating. The syndrome causes diarrhea and great discomfort. After consuming what Kansans crudely called dingle berries, these beetles enter the rectum, when opportunity presents itself for more of the same. Because of hooks and barbs on their forelegs it takes a physician to dislodge them. The spines, horns and other paraphernalia used to tunnel forward prevent easy retreat.

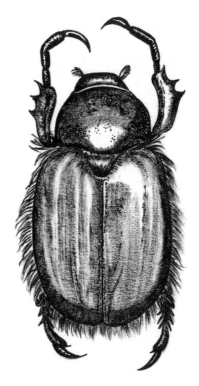

Pseudocotalpa giulianii *is a sand-loving scarab often endangered by all-terrain vehicles in fragile country. This bug is a pretty Palomino gold, and has a pale, flax-colored fringe around its shell.*

17
Scanning the Literature

After scanning enough peer-reviewed literature to choke a herd of horses, I've concluded that dung beetles have commanded more scholarship than some medical disciplines, or lives of Napoleon, or tabloid pages on aliens from outer space.

With the able assistance of a University of Virginia librarian I obtained a list of peer-reviewed publications that can't possibly be complete. I counted the titles of some 50 books, most of which seemed unavailable through inter-library loan, 30 Ph.D. dissertations that might be available at a reproduction cost of between $50 and $150 each, and 1,420 scientific articles published between 1973 and 2005, some in journals so difficult to find that they rate in value with artifacts under glass in a museum. Each of those examined contained a morsel of information fleshed out in excruciating detail, whether the article delineated the powerful logic of the experiment or simply described the anatomy of a dung beetle trap.

Scanning and reading enough pages to tire out a computer's voice program, I managed to learn that dung beetles sometimes transport, often quite a distance, seeds still viable after being passed through the bovine digestive tract. During the 1970s at least, cows on pasture pelted the grass with a million evacuations a day, some larger than others. Dung beetles are as territorial as bureaucrats. They fight, but they also live in peace. They tolerate, yet they can exhibit the intolerance of a professor too erudite to bother with teaching. This is not to suggest that much of the research seems frivolous, which it is, but to give some idea of what all these investigators are all about.

In one manuscript that came to my attention, Truman Fincher said, "Scientists today are doing research that only benefits themselves. They want to get patents on what they develop so they can become rich, or they do simple research projects each year to get enough data to publish a little manuscript to justify their job. Some of you may have heard that most research data published by scientists today is not worth the paper it's written on."

Fincher's indictment may sound harsh, but it also explains all the prolix articles on dung beetles and the paucity of interest in field work, farm and ranch colonization, and actually repopulating America with those valuable guest workers. "I was discussing this with a friend," Fincher once said. "He made the comment that if 80% of all the agricultural research scientists were fired today, no one would ever miss them because they have not and never will accomplish anything to benefit the American taxpayer."

Truman took pardonable pride in his work because when he walked across a pasture he could see research results. He looked to the day when Southern pastures repopulated by foreign imports cleared away manure the way dung beetles cleared away elephant dung in Africa.

Fincher and his associates, in fact, carried out some rather sophisticated investigations, one of which I would like to set forth here. Dung from steers injected subcutaneously with the parasiticide moxidectin (0.2 mg/kg of body weight) was bio-assayed with the dung beetles *Euoniticellus intermedius* and *Onthophagus gazella.* There were no effects on the mean numbers of brood balls produced by either beetle species or on the emergence of adult beetles from brood balls constructed from the dung of injected cattle. The sex ratio of the progeny of both species was not affected by moxidectin.

If you can frame the question and have the patience of Job, you can probably find an answer in the scientific literature published since 1971. A vast body of reports also memorializes scientific efforts from the 1920s onward, and a few tomes even go back to pre-American Revolution days.

It has always been the objective of chemical companies to enlarge their tyranny over biology. Accordingly, experiments proving or disproving the efficacy of various chemicals ran like a jagged scar through the grand dung beetle experiment for over a quarter of a century. Some chemicals allow dung beetles permission for life. More do not. With thousands of compounds named in the *Farm Chemical Handbook,* I hesitate to wonder how many have contributed to the virtual annihilation of dung beetles previously serving the nation's pastures.

Readers are entitled to a few samples that may not have been covered elsewhere in this text. Of the many lists ordered up and examined, many offered a concise synopsis or precis, a few of which are quoted verbatim.

Who steals the eggs? *Coprophanaeus telamon* buries decomposing eggs in western Amazonian rain forest (Coleoptera: Scarabaeidae), by A. Pfrommer and F.T. Krell in the *Coleopterists Bulletin.*

"On 25 December, 1995, 80 hen eggs were deposited along 4 transects of 500 m length in terra firme rain forest in Ecuador. After 26 days, 20 eggs were found buried. A single male *Coprophanaeus telamon* was found in a cavity directly below a buried egg. A tunnel with a diameter of approximately the body width of *Coprophanaeus telamon* met this cavity from the side. Similarly, a single female was found at another buried egg. These 2 buried eggs were intact, but showed soft parts of the shell where the calcium was apparently dissolving. At all other sites where buried, decomposing eggs were found, the same cavity and tunnel were observed, indicating *Coprophanaeus telamon* activity. We suppose that the volatiles 2-butanone, cresol, indole, skatole and butyric acid are responsible for attracting Coprophanaeus because these components of dung odor attract dung beetles and are also present in rotten eggs. A number of them are also produced by bacterial spoilage of raw meat. Abandoned clutches, infertile eggs or eggs with dead embryos of ground-nesting birds may be used as a resource by dung beetles."

Now consider this one authored by D.M. Spratt from the *International Journal of Parasitology.*

"The broad spectrum and high efficacy of the macrocyclic lactones against nematode and arthropod parasites of livestock and companion animals are unprecedented. To varying degrees, cattle, horse, sheep, pigs, and dogs are all utilised by humans for economic gain. Detrimental impact upon non-target animals is considered acceptable in eradicating parasites because of their economic importance to commercial livestock production. Production will increase when these parasites are eliminated, but we remain oblivious to the long-term consequences of our actions. What are the economical limits to rural economic activities? Decomposing animal faeces help to maintain our ecosystem by returning valuable nutrients to the soil. Dung fauna — fungi, yeast, bacteria, nematodes, insects, and earthworms — play a nonconspicuous but important and varied role in this decomposition process, a role dependent upon many factors, especially environmental ones. Anthelmintics and pesticides are of considerable value in agriculture, but largely at an unevaluated cost to the greater environment. We have insufficient knowledge of the extent to which a spectrum of anthelmintics and pesticides affect ecological function and ecosystem resilience in our commercial plant and animal production systems. It is time we developed a genuine interest in avoiding "the dialogue of the deaf" that in the past has minimised interdisciplinary research between environmental ecology and commercial plant and animal production."

Here's one from the journal *Pedobiologia* by J.P. Lumaret and N. Kadiri: *The influence of the first wave of colonizing insects on cattle dung dispersal.*

"The primary and secondary effects of the exclusion for one month of dung beetles from fresh dung pats in pastures were studied in field experiments in southern France. The evolution of two dung-pat series placed monthly in the field was followed for four years: pats open to colonization by insects; pats covered with wire gauze excluding beetles for one month. The total disappearance of exposed pats ranged from 2 to 4 years, with a high variability among series. Covered pats took 1.7 to 2.2 times longer than exposed pats to disappear completely. The influence of beetles is essential during the first month after the dung is deposited, though burying dung in tunnels weakens the dung pats which subsequently more easily disintegrate. A multiple stepwise linear regression was used to test the joint effect of six climatic variables on the rate of decay. According to the season, two or more variables could explain a large part of the breakdown of exposed pats. For covered pats, climatic factors were of little influence. The consequences for the environment due to the difference in decay rate between exposed and covered pats are discussed. The evolution of protected pats corresponds to some characteristics observed when dung pats are polluted by toxic worming drugs given to cattle, resulting in a slowing down of recycling processes."

No breed or dung style is overlooked. Sheep dung, often ignored by cattle-oriented workers, is dealt with nevertheless. Briefly, here is a synopsis of an article in *Veterinary Parasitology* by K.L. King entitled "Methods for Assessing the Impact of Avermectins on the Decomposer Community of Sheep Pastures."

"This paper outlines methods which can be used in the field assessment of potentially toxic chemicals such as the avermectins. The procedures focus on measuring the effects of the drug on decomposer organisms and the nutrient cycling process in pastures grazed by sheep. Measurements of decomposer activity are described along with methods for determining dry and organic matter loss and mineral loss from dung to the underlying soil. Sampling methods for both micro- and macro-vertebrates are discussed along with determination of the percentage infection of plant roots with vesicular-arbuscular mycorrhizal fungi. An integrated sampling unit for assessing the ecotoxicity of avermectin in pastures grazed by sheep is presented."

Some of the papers are short, a page or two. Others bark at the heels of *War and Peace*. Manure is almost always the focal point, dung beetles the actors.

Here's a synopsis from the *Bulletin d'Écologie* entitled "Effect of Excessive Accumulation of Manure on Soil Arthropods in Grazing Areas" by J.P. Lumaret and M. Bertrand.

"The authors analyze and compare the biological effects induced by a limited deposit of sheep faeces (single pads) or by a permanent accumulation of manure in sheepfolds under Mediterranean climatic conditions. The burying of single pads by dung beetles strongly stimulates micro-arthropod populations (crowding effect) and induces a rapid resorption of this resource by the edaphic system. An excessive accumulation of dung in the fold chokes up the system: the resource (fecal material) is only exploited by a few species of saprocoprophagous beetles, while the edaphic populations of mites are strongly depressed. In spite of important insect densities, low manure consumption explains the perenniality of the dung layers."

This one's from the "Proceedings of the 3rd Australasian Conference on Grassland Invertebrate Ecology" by T.J.R. Smith and A.A. Kirk. It's entitled "Dung Beetles and Dispersal of Cattle Dung."

"The disappearance of cattle droppings deposited about 1 August was followed in pastures at an experimental farm in the Copenhagen region of Denmark in 1974, 1975, and 1977. Pats covered during the first 5-7 nights to exclude night-flying beetles such as *Aphodius rufipes* were compared with freely exposed controls. The night covering considerably depressed rates of decay. The disappearance times for 75% of covered and uncovered pats being 54 and 32 days, respectively, in 1974 and 64 and 42 days in 1977. In the hot, dry summer of 1975, when earthworms were not active on the surface, only 35% of the dung disappeared in 2 months, whether the pats were covered or not. It was determined that larvae of *Aphodius rufipes* were responsible for the loss of 14-20% of the dung and earthworms for about 50%. It appeared that the activity of Aphodius and their larvae accelerated the aggregation of worms under the pats, thereby promoting the rapid disappearance of the dung."

Here's one from the *Canadian Journal of Plant Science* by A. Macqueen and B.P. Beirne entitled "Effects of Cattle Dung and Dung Beetle Activity on Growth of Beardless Wheatgrass in British Columbia."

"The nutrient value of fresh cattle dung for British Columbia range-land plants was evaluated over two seasons in a pot experiment using depleted range soil and beardless wheatgrass, *Agropyron spicatum*. Dung treatments comprised fresh cattle feces (200 g) mixed manually with the soil, or worked in by the dung beetle *Onthophagus nuchicornis*, or placed intact on the soil without burial. There were also fertilizer treatments at 67 and 269 kg N/ha. Total incorporation of the fresh dung into the soil increased total crude protein production, potential seed production, and the vigor of the grass over a 2-year period. Burial of an average of 37% of the available dung by beetles caused a 38% increase in crude protein over that of the control. Crude protein production of grass treated with 67 kg N/ha was 95% higher than the control, while that treated with 269 kg N/ha was 144% higher. These rangelands have an impoverished native dung beetle fauna in comparison with some other climatically similar areas of the world. Efficient dung beetles should bury cattle dung more effectively during the growing season than do the present species."

You say, "Enough already!" Enough already, indeed. I have merely constructed the reason for creating still another book on dung beetles. The manual published by Australia's CSIRO is hopelessly out of print, and the door-stopper books on the subject often take an Act of Congress to dislodge, an exercise hardly one cattleman in the country has time for.

There is an excellent *Texas Bug Book* by Malcolm Beck and Howard Garrett from Texas University Press, but its coverage of dung beetles is limited to about three pages. I found a copy of *An Inordinate Fondness for Beetles* by Arthur V. Evans and Charles L. Bellamy with photos by Lisa Charles Watson in a used book store. This University of California offering doesn't trouble itself too much with italic nomenclature and tells a straightforward story of beetles with dung beetles in an almost non-existent role.

Having exhausted an inch-high list of learned prose and reading the footnotes, one learns of the chemical companies that support research, supply the nostrums, and, sub rosa, order up results or invoke contract provisions, then shred the experimental results. Mention of a product in another footnote or addendum informs with measured sincerity a lack of endorsement or recommendation.

Rarely does the literature postulate an overarching economics, such as the value of dung beetles to the red meat industry. That kind of in-depth style of projections fell to Truman Fincher, who discerned the "greatest invention" still to come and the rancher's need for orientation on what really builds the bottom line.

Entomology and parasitology are each horns of the same dilemma, one discipline relies on the other. Thus, the spate of reports that makes the connection, usually with the horn fly, buffalo fly or face fly in the crosshairs. The notes that lace the rest of these pages together have generally emerged from this difficult throne of esoteric knowledge.

Here is an abstract from one College Station experiment that may be of maximum interest, one of the countless papers bearing the byline of George T. Fincher and interested associates, as the title makes clear in "The Survival of Horn Flies According to the Time of Dung Deposit by Cattle," published by the *Southwestern Entomologist*. Footnotes give us the intelligence that some of the writers are connected with Rhone Merieux, Inc. and Rubin/Haas Company. I have not burdened these pages with such information as a rule, the information being of no great importance except to reveal the guiding hand behind much research. The quotation sets out the scope of the research.

"Cowpats were flagged after deposition on pasture at 0500, 1100, 1700, and 2300 hours and exposed to natural enemies for 6 hours or 6 days. 61% of the horn flies emerging from pats exposed for 6 hours emerged from cowpats dropped at 2300 h and exposed between 2300 and 0500 h when little or no flight activity by natural enemies occurred. 15% of the horn flies trapped from cowpats exposed 6 hours emerged from pats dropped at 0500 h, 14% from pats dropped at 1100 h, and 10% from pats deposited at 1700 h. From cowpats exposed to natural enemies for 6 days, 52% of the horn flies emerged from pats deposited at 2300 h, 23% from pats dropped at 0500 h, 17% from 1100 h deposits, and 8% from 1700 h deposits. Mean numbers of dung beetles, predator beetles, and flies other than horn flies obtained from cowpats after both periods of exposure are discussed.

"The horn fly, *Haematobia irritans,* is an obligate blood-sucking ecto-parasite of cattle, causing annual losses in excess of $876 million in the beef cattle and dairy industries of the United States (Kunz et al. 1991). Both sexes of this pest remain on the host; if provided with an oviposition medium, females oviposit any time of the day or night (Kunz et al. 1970). Thus far, insecticides have been the only method of managed control and, as with many insect pests, resistance has developed (Sheppard 1983, 1984, Kunz and Schmidt 1985, Sparks et al. 1985). Also, there has been increased pressure from government regulatory agencies and the public to reduce the use of chemicals for pest control and to use only insecticides that are short-lived to avoid contamination of food, water, and the environment. These events have generated increased interest in the use of biological agents to control many insect pests."

Biological control of insects has long enjoyed "flag up" support at the USDA. This statement seems to be denied by their attitude regarding dung beetles. The screwworm program comes to mind, circa the 1950s. The program commanded a great deal of attention in *Veterinary Medicine* magazine, where I once upon a time handled an editorial pencil. Screwworms were hatched and sterilized on the island of Curaçao, then released to mate over the southern states. The program was a great success and promised more of that type of research covering other questions in other areas.

The devastation of the blood-sucking horn fly invited a scrutiny only SS PRP (statistically significant, peer-review publication) could account for. For three decades, the literature flashed go-ahead signals, but the makers of toxic, genetic chemicals were ever in the wings. A tantalus of chemicals asked for research and the asking was almost always accompanied by money.

Still, there is an honesty and a statement for academic freedom in almost all of the papers I ordered up and read. Thus we find extraordinary work on visual systems, moon navigation, nocturnal habits, co-existence with worms, ants and other insects, and anatomies in color schemes that put the rainbow to shame.

"But we know most of these things," *Texas Bug Book* author Malcolm Beck told me. He was visiting the West Ranch at Ozona, Texas, a Holistic Resource Management operation. HRM in fact manages the ranch with Allan Savory, the often present adviser. "Joe Maddox is the working manager," said Beck. "It was near a windmill by a water trough that I made an observation. The ground was stomped, eliminating the grass, but the soil was excellent and there was plenty of manure.

"I'd never seen single tumblebugs before. Usually they work in pairs. The females were cutting out the balls and rolling them, pushing backwards, uphill to a burial ground. The area was full of rocks. Those bugs had chosen their spot near a grove of trees where the soil was deeper."

The balls were as large, or larger than, the bugs themselves. Malcolm felt sorry for them. He almost wanted to carry the balls up for the tumblebugs.

"They'd start up the hill, then they'd run into an obstacle. So they'd jump up on top and look around just like Hawkeye in those Fenimore Cooper stories. Then they'd pick out a path and go to work again. It wasn't the work of Sisyphus. The ball never rolled downhill again. It was always installed into a soil chamber."

Beck became quizzical. "You know, there's a lesson here. Rollers work in pairs on brood balls only, but here was a single female going solo on a feeding ball. The scarab tumblebug is the only one I've ever seen and I judge that's the case with most people. Kids sure like to watch them."

They do. The tunneler remains anonymous and the dweller is a little fellow who homesteads and stays out of sight. But the tumblebug is a showman, a veritable Barnum and a Houdini of strength. Malcolm concluded, "So much research and authors and those Latin names go in one ear and out the other but the observations you make stay with you."

There is probably research in those 1,400+ articles between 1971 and the present that address the matter of parasites on dung beetles. Beck tells of them in *Texas Bug Book*. In the Sand Hills, Beck found a green beetle and he discovered a parasite that lives on the back of the roller. It's a very small gray insect that hangs on in clusters. They live off fragments of manure. When the tumblebug proceeds to fly, they vacate their residence on the wing covers. Like the bank robbers that J. Frank Dobie wrote about, the most successful require anonymity. The notorious are the failures.

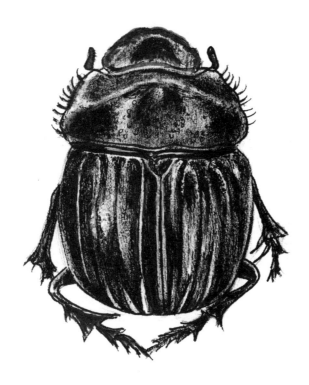

Dichotomius carolinus *is a dung beetle with smooth wings and light texture on its forequarters. Its stellar role reclaims its status as the largest native dung beetle in our quest for control.*

18
Competition

There is an interplay in Nature that would tax the biggest computer even if it could be programmed. There are nematodes that hitch a ride on beetles to go from one batch to the next. The evolution questions posed are beyond instant comprehension. Even though of no great consequence to dung beetles, when dung beetles donate their service, they suppress the parasitic interlopers.

Our aside expressed above calls into question the cow pie as the prince of nutrients. It is and it isn't. There are beetles that live on decayed mushrooms and forest organic matter. Some bite off wood. Years and eons of time have intervened to direct the beetles of which we speak to the food sources they enjoy, indeed, can't live without. Beetles with soft mandibles like to slurp, incisors straining out the undigestibles. As we view maceration and digestion, we make clear the five-star dining requirements, beetle to beetle. Beetles with soft mandibles require soft food, those with hard mandibles prefer hard food.

The rancher who wants the dung removed from his pastures sees such complexity and no longer wonders why so many researchers get lost in the horse latitudes of classification, taxa, anatomy, whatever and forget all about bringing suitable dung beetles to the pasture for the cleanup chore.

Hobby scholars can take nocturnal species apart and put them back together again. Those that fly in twilight or full daylight have been categorized, mounted, labeled, described and tagged with Linnaeus-type names until the fount of knowledge takes on the color of baseball statistics going back at least to Abner Doubleday. The topics are partial to nocturnal flyers. Field observations made in African savannas and Central American tropics now have SS PRP status, legitimized by the rubrics of settled science. The common denominator is the efficiency of excrement exploitation, a food so perishable it reminds us of lunch meat in the refrigerator under attack of Listeria and Salmonella in spite of refrigeration. If microbes can live and thrive in hot vats on the ocean floor, then it is reasonable to expect microorganisms to preempt dung beetles when remanded to soil burial.

We applaud the rescue of a valuable resource of prairie, savanna, veld and steppe from the biological and climatic disturbances ever afloat, and we thank the dung beetle the way a comic responds to a hiker, "Tired of walking? Run a while." But we fail when we walk away asking our folly to repair itself.

Nocturnal insects handle bright sunlight badly. At dusk, the valued dung release may be hard to see for a rural boy chasing a fly ball in a pasture, but the dung beetle is equal to the task. Dung beetles use the polarization pattern formed around the setting sun. When the sun drops 18 degrees beyond the horizon, the pattern is lost. The moon, however, creates a new polarized pattern in the sky. It lasts as long as the moon's orbit permits light on the area involved. Under moonlight, beetles roll their balls, send out scouting squadrons, and — as the Brits say — "carry on," but an overcast sky cancels out much of the navigation instinct or vision.

One of the announced goals of the USDA dung beetle project housed at College Station, Texas was international cooperation. As an adjunct to this goal, Canada promptly provided the intelligence that some 450 species of insects can be found in cattle dung. All had scientific names that could march by for a quarter of an hour. Some of these species — stable flies, horn flies, face flies — all important pests of livestock, received a great deal of attention by the government's biological control people. As an example, there is a common beetle in Canada that feeds on the eggs of pestiferous flies. As in the U.S., many valuable species have drawn little attention. Yet, Nature has decreed that the proliferation of parasites that gravitate to the dung pile should have a balancing counterpart. From this arrangement flows economic value if the animal husbandry specialist has the wit to recognize it.

Truman Fincher's 1981 paper entitled *The Potential Value of Dung Beetles in Pastures and Eco-Systems* identifies several ways in which degradation of dung returns pasture to grazing after being strangled by dung. Foremost was the recycling of dung — sheet composting, if you will — into the soil. This payload can be calculated in terms of tons. The nitrogen factor has received lavish attention before. Not so with parasites. Beetles feeding in the dung mass cause it to dry out quickly. It is this business that disrupts the life cycle of helminth parasites attempting to hatch in the pliable excreta. The reduction of parasites in cattle can be both imagined and calculated. Further, reduction of pest flies, whether considered parasites or not because of desiccated dung, cancels out breeding.

In 1981, Fincher calculated the value of services that can be credited to the little bugs that favor dung. His numbers will be enlarged upon in the next chapter. Increased grazing and nitrogen salvage come to over $603 million for the area being calculated, a not insignificant sum. Reduction of cattle parasites fleshes out that calculated bottom line by something less than $200 million.

The much yearned-for greatest invention postulated in the first chapter has still to be refined with an eye to physics, biology, nutrition, reproduction and a few dozen causes of the cause. The incorporation of organic matter into the soil is a value not even Fletcher Sims's Mechanical Scarab can salvage. Indeed, real accounting demolishes the university model of concentrated feeding, rescue chemistry and hasty termination of fat cattle with genetically modified grain, Arm and Hammer bicarbonate of soda, protein bypass and the feedlot bloat-tube down the hapless critter's throat.

Yes, organic matter connects. It enables water retention, as we have seen, and it preserves soil viability, and, not least, it reduces or removes the livestock pests that homestead cattle manure. The disease problem — read that, the veterinary bill — lives on dung beetle depopulation. Horn flies, stable flies, face flies, all styles of dung-feeding flies suggest a new look at arthropods that dine on pests.

The chant that "we have chemicals that do a better job" or "better living through chemistry" emanates from the devil's pantry, always without an accounting statement quantifying the deficits, including the final bill called Nature's revulsion.

The greatest invention, real-value bookkeeping, will have to rely less on computers and move on to a course of study all but excised from college curriculums, logic! Each year, insecticides and the labor incident to their application sustain an industry with funds that rightfully belong to the primary producer. This industry cancels out Nature's own and imposes deficits that cannot be recovered under present public economic policy. The horn flies, face flies and countless parasites seem to smile, proliferate and torment cattle, each bite, each blood-suck tolling the bell for the profit sheet. Many of these parasites are imports that arrived without their natural predators in tow.

During a few weeks of the year, southern climate, temperature and other conditions preside over a horn fly hatch. Depopulation of dung beetles due to inept pesticide intervention and inconsistent control determine the ecology for the year. One error allows a buildup for the next blunder. It became the finding at College Station, Texas that the lack of predator activity, more than any other factor, made possible the advanced tyranny of the horn fly.

Beetles are not alone in battling the cowman's manure. There is a small wasp that is known to parasitize some stages of horn and face flies in the cow pie. Unfortunately, it is seldom able to make much of a dent in the fly population. It is now a matter of record that some 22 species of parasitic wasps have been identified as agents capable of assisting the dung beetle in the mortal combat arena. The jury stays out because the wasps seem to prefer parasitizing fly species other than those targeted by the goals of bio-control. Favored parasites have been released through the years. Hardly any seem to have the potential of the 15 species of dung beetles field-colonized under the program described herein.

The public learns of these things slowly. The metro dailies spot a short feature now and then, but the public walks away with little more than a precis of what the sacred scarab and its kind are all about.

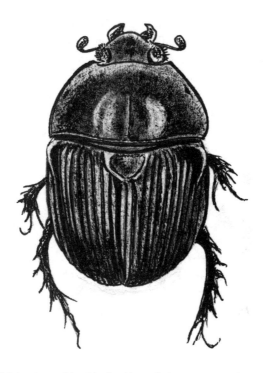

Geotrupes egeriei *is almost blue-black with a slight texture on the shell. Its wings are deeply channeled. A truly American native.*

19

The Greatest
Invention Revisited

It may be that the greatest invention for century 21 could be accounting. We've now had our turn with the double-entry accounting of medieval monks, and we've endured the creative refinement thereof by Enron, Tyco, Global Crossing, and — that word again — those human klepto-parasites, *yclept* public servants. The inability to keep books has enabled Congress to increase its salaries incrementally each recent year, yet deny minimum-wage workers the alleged benefits of the culture of cheapness. The same accounting system has raised Congressional salaries from $10,000 in 1947 to $165,000 at the middle of the first decade of century 21.

I make short reference to Oppenheimer's *Cowboy Arithmetic* partially because the sprawling enterprise went broke, but mostly because the bookkeeping of those cowmen ignored real costs, real physical possibilities and the real debauchery of agriculture. In the middle of all this was and remains the dung beeetle.

Perhaps the author of the invention still to be made will return to the technocrats of the Great Depression era who were always talking about ergs and joules, or the physicists who have everything denoted in atoms, electrons and protons. For now, we have to deal with dollars, a fact that Truman Fincher was early in comprehending, real costs and real savings independent of supply and demand rubrics, packer and stockyard manipulation and errant public policy.

The dung beetle, armed with its DNA directives, simply abides. Would the incidence of various animal diseases be lowered by the introduction of proved species? Would a reduction of parasite infestation be extended to wildlife? The case for feces removal from pastures stands on its own feet. Putting numbers with reduced cost of producing animals on pasture is challenged routinely by schoolmen who conjured up the confinement feeding system with its resultant ills, not the least of which is a wide separation of the Omega 3 to Omega 6 fat ratio, feedlot cattle often exhibiting a 1 to 20 ratio, not the 1 to 1 required by Nature. Dung beetles cannot function in a feedlot where manure is forever hammered into slabs as hard as a city street.

Soil's requirement for organic matter is well known, yet it is possible to consult a library full of soil texts without a single reference to Nature's very own compost maker. Feces on top of the soil makes poor compost, it wastes nitrogen and phosphate, and in general squanders the economic value inherent in manure. The service that dung beetles perform by feeding microorganisms is amazing. Still, this factor can be computed based on the subtraction symbol. Chemicals and chemical residues of man-made molecules inhibit and destroy dung beetle colonies. They can sometimes be neutralized by beetles, but at ruinous expense.

Suppose the accounting system permitted calculations for value destroyed at a packing plant — excess trim and everything from waste to the pre-inspection art of adding sodium benzoate to tainted meat. The greatest invention will ask for an assist from parasitologists, entomologists, ecologists, agronomists and taxonomists. In an agriculture distorted by bad science, the greatest invention will require immigrants, not guest workers.

American investigators and officials are risk-avoidance prone. Decrees from insurance companies filtered by lawmakers have removed high diving boards from swimming pools, even though pool accidents are next to non-existent compared to walking across the street. The late great dung beetle program was slowed to a crawl by precautionary overkill because a beetle was imported. Other nations came to College Station, scooped up beetle eggs and took them back home, no fuss. How does the paper drill compute in larding out expenses that not even a colony of gazellas can cancel out? Larval migration from 11-15 centimeters underground is impossible, the dung beetle tells us. Such a disclosure may marginalize the legendary tumblebug, but not the tunneler.

Onthophagus taurus showed up in Florida one day like an immigrant at Ellis Island, WOP — without papers. This was in 1971. The little creature apparently hitched a ride with a cow or a zoo animal, or perhaps arrived as a stowaway shrouded in secrecy.

We expect to grade efficiency in terms of dollars, but dollar values are ephemeral in an age of inflation. Numbers for one decade seem ancient a decade later. The role of dung beetles in nutrient cycling invites value calculations in denominations that can't be manipulated by the absconders of high finance. Even in a vacillating world of money numbers, the primitive calculations involving nitrogen and phosphorus set off a bang. There have been studies aplenty, for which reason I feel secure in revealing the metes and bounds of nutrient extraction, nutrient waste, even nutrient value. The data base says that for every 100 head of cattle, there is a potential loss of 851 pounds of nitrogen and 237 pounds of phosphorus. That is a physical reality. The dollar value depends on the year one makes the extension. Clearly, the nutrient loss occurs when dung remains unburied and subject to leaching and weathering. In any case, that loss amounts to 20-24% of the deposited dung. Burial allows incorporation of nutrients where the plants feed.

The business of so much nitrogen, organic or inorganic, lost to the ambient air has ramifications never discussed in or out of any connection with dung beetles. Anhydrous ammonia is largely lost while being knifed into the soil. So is a great deal of the three-number fertilizer with N as the lead number. Those one billion cow patties exhale nitrogen quite rapidly if not taken underground promptly. Forgotten in even an Al Gore treatment of the subject is the fact that nitrous oxide is a greenhouse gas on par with or exceeding carbon dioxide. Nitrous oxide is 183 to 212 times more effective than carbon dioxide in toasting the environment. It doesn't take a physicist to make a defensible "greatest invention" calculation validating this forgotten pollutant and its errant nitrogen origin. It could well be that the conventional use of agricultural nitrogen in a non-organic matrix is hard on the heels of blame-heavy carbon dioxide. As public policy continues to drive cattlemen and row-crop farmers off the land, the hands that remain turn to heavy nitrogen, and feedlot animals have no chance at all for dung beetles to take this nitrogen treasure-trove underground.

In order to flesh out this nitrogen-cowpattie-rowcrop farming connection, it is necessary to recall a little basic chemistry. Nitrous oxide, N_2O, is quite similar in molecular weight to carbon dioxide, CO_2. This knowledge requires us to measure carbon dioxide against nitrous oxide as greenhouse gasses.

Nitrous oxide is created by the composition of nitrogen fertilizer. Anhydrous puts out 2-3 times more nitrous oxide than basic urea or ammonium nitrate. Nitrogen contained in fresh cow dung takes its place among the pollution factors if left to the desiccation process. The most stable molecule that nitrogen accounts for, N_2, and the oxygen atom in nitrous oxide, N_2O, is easily lost to produce free nitrogen. The free radical oxygen atom is then involved in atmospheric chain reactions. Nitrogen dioxide, nitrous oxide, and other nitrogen oxygen gasses are now called NOX, a sin of "emission" from internal combustion engines. Ozone is O_3. Oxygen is O_2. Ozone easily sheds an oxygen atom, which becomes a free radical starting a chain-reaction in the atmosphere. The chief reason for no-till farming is the restrained release of carbon dioxide and nitrous oxide, ergo less greenhouse gasses into the atmosphere.

The above is quite a load to put on dung beetles, a bug so insignificant that no one bothers to issue a grant so that its genetics can be fully studied. We can compute the emissions from automobiles and coal-fired industry. With data on nitrogen use not a part of the natural nitrogen cycle, the culprit may well sleep in our midst — calm, tranquil and uncalculated. To all this the dung beetle must shrug a shoulder and go about the business of sequestering the nitrogen in its favored food.

We marvel at the simplicity of the computations once we adjust the thinking that must attend the proposed "greatest invention." We blanch when we compare organic nutrients to their industrial equivalents, but we proceed nevertheless. That unburied dung from 100 head of cattle equals 0.8 tons of urea and at least 1-2 tons of single super-phosphate per annum. The arithmetic plows through all the obfuscation that chemical purveyors have to offer. Simple arithmetic requires us to consider pesticides necessary to control the profit thieves that dung beetles would have controlled in the first place while quietly fulfilling their soil-nurturing duties. Any adjustment of attitude points to the cost-effective business called field colonization.

CSIRO, the Australian agency, had its moment in the sun before economy measures closed down supported dung beetle colonization. The thinking is correct, the scientists said, but the numbers assigned to the fact were rock-bottom conservative. The Australians figure that a good dung beetle population eliminates fully 15% of the pollution ready to enter streams. John Feehan questions the low official computation for reasons he demonstrates to school children with a bucket of water.

No one would set fire to an asset such as a tool shed or a barn. Yet that is what many farmers do with drenches, glyphosate by any of 30-40 names, Roundup included. Drenches take out beetle colonies the way Indians once took out a 7th Cavalry regiment, to the man, or in this case, to the last bug. Unfortunately, or fortunately, Nature has her own stern code. John Feehan reports seeing foxes eating large winter-active dung beetles like so much ice cream.

It is axiomatic that science is largely the process of asking the right questions. This ability comes with education and life's experience, with heavy emphasis on life's experience and the culture in which one lives. Total acceptance of theory-period instruction puts metamorphosis into reverse and turns butterflies into slugs. The slugs that spend billions on destruction and spare not one niggardly dime for a science that has the potential of returning to agriculture its natural mandate for fecundity and balance are as sightless as have been all the blunderers of history. The balance sheet beckons. It asks and answers everything that can be known about the eternal equation. Having overlooked nothing, the greatest invention will be required to answer everything.

The greatest invention that is still to come has its work cut out for it. Klepto-parasites are not the only thieves in Nature. Mismanagement of manure is a thief. Failure to understand the fine points that examination of Nature has revealed steals from the bottom line and is never called to account. The spreadsheet that a cowman has to consider may not be printed yet, but Truman Fincher has proposed some of the items. These items govern. The astute reader will have assembled some of them on his or her own from the foregoing narrative. It is both a review and an extrapolation of an untold story and a citation that academia failed to issue.

Dung beetles have names, family, genus and species. The family name is Scarabaeidae with some 30,000 species. Including other dung dwellers from the Hydrophilidae, Histeridae, Staphylinidae and other families would raise the species figure dramatically. American species have all been classified, dissected, analyzed and showered with wonderment, the biggest wonder of all being the indifference that ranchers and scientists have accorded this most valued creature. It would take an encyclopedia the size of an unabridged dictionary to get into the bug classification game. Suffice it to note that Fincher and colleagues finally settled on a few species for repopulation of North American pastures, some of which have been illustrated at the head of each of these chapters.

The value of dung beetles to pastures and the ecosystem was first acknowledged by George Bornemissza, the native Hungarian. Wrote Fincher, "The viability of every pasture eco-system is based on the normal function of its nutrient cycle." The grazing animal is one component. Its role in keeping the system operative is acknowledged by all who think first of ecology before they turn to salvage via chemistry. In effect, the bovine animal is its own manager of its environment if allowed to function with an ample supply of air, water and forage presented as living room. Mismanagement brings on repercussions.

Bornemissza's message bears repeating. "The Australian grassland eco-systems were profoundly disturbed by the arrival of domestic stock. Prior to that, the nutrient cycle is thought to have functioned smoothly, and that the dung of the principle herbivores, the marsupials, was relatively unimportant and probably never accumulated in polluting quantities. The portion of the marsupial dung was buried by the native dung beetles, and this burial sped up decomposition and returned essential nutrients to the soil. The introduction and rapid increase of the number of domestic grazing animals progressively upset this primitive cycling. As the droppings of horses and cattle accumulated, the pollution problem arose in many areas — to the extensive fouling of valuable pasture land."

In the U.S., when buffalo passed over an area, or deer and elk gifted their droppings — as did birds and smaller animals — native dung beetles and other coprophagic organisms easily handled their dung. The beetles easily blended their function to the demand of nature. They removed the droppings from the soil's surface with ease and efficiency.

Over the past 60 years, the near-universal use of salt fertilizers and toxic rescue chemistry has turned the "American Serengeti" into a dung beetle desert. New grasses and greater yields have increased carrying capacities and therefore a new inventory of dung awaiting disposal. The dung is ready for transport underground the minute it hits the turf. Unfortunately that is not what often happens. Cows refuse to graze spoiled grass and spoilage takes place over a wider area than the cow platter, an area often as big as baseball's home plate. With beetle population all but gone, cow pads often stay put for months.

Speaking before a San Antonio assembly of *Acres U.S.A.* readers, Truman Fincher mused about the apparent and real decline of the beetle population in Georgia, "Many older farmers in Georgia have stated to me that there are not as many tumblebugs today as there were when they were growing up. However, studies of the dung beetle fauna on the coastal estuary of Georgia reveal that thousands of beetles can be captured within a few days, including tumblebugs. I was amazed at the number of pairs of *Canthon pilularius* that I observed daily on Ossabaw Island at a single observation point. They were rolling balls of cattle dung in an open area of an acre. After 12 years of trapping beetles in Georgia, I have never seen this many tumblebugs working at the same time in one place."

Since no crops are produced on the Georgia islands, non-use of farm chemicals has been the norm. Moreover, area livestock are not treated for parasites or pests. New forages, chemical insecticides, intensive grazing and some few modern husbandry practices all have conspired to depopulate dung beetles on the mainland. Clearly, an overload of undigested manure on pastures reduces the available grass, and this translates into ever-faltering figures on the bottom line. Many of these figures have trailed their way through these pages in order to capture the interest of the cowman subliminally while his tutorial on dung beetles proceeded.

During his more than two decades as one of the nation's leading dung beetle aficionados, Truman Fincher made countless calculations on acres recovered from grazing and economic losses, such as nitrogen lost when dung isn't melted away, captured by beetles or removed. Rank growth around each cow platter costs as much as the area covered with dung. Horn flies and face flies move in — rather they breed in fresh dung. Undisturbed, the cow pie becomes a residence for various stages of gastro-intestinal parasites that torment and debilitate livestock. The nitrogen content in the cow dung goes up, not in smoke, but as an unseen miasma.

Other nutrients, research has revealed, are taken up for months, even years, and not available to feed the organisms that feed the pests.

Grazing animals at a stocking rate designed to consume the available forage that fertilization is capable of yielding courts the danger of poor animal performance. Some cattle producers have sought to make an end-run around the surplus forage problem by making pellets, stacking hay and otherwise servicing the feedlot. In Nature, the world energy situation has endured a paradigm shift. Cattle raising that depends on expensive machinery and fossil-fuel energy consumption is no longer defensible, on paper or in fact. Thus, the beetle comes back in focus, and feces removal becomes a technology devoutly to be wished. Harvesting and pelletizing forage is too expensive and too unscientific nutritionally.

Nature's first sanitary engineers were the dung beetles. These dung rollers have biographies that roll forward through climate changes, volcanic upheavals, even strikes of meteorites from outer space. Likely, North America never had nor could have had the proliferation common to tropical Africa or South America. If snow and ice pushed mammals north from the south, and south from the north, it's certain that the beetles followed their source of food.

During his tenure at College Station, Texas, Truman Fincher made several counts, a few of which are of interest to North American graziers. Data available to the researcher revealed approximately 1,137 species of Scarabaeidae in the Western hemisphere. 945 live in South America, 157 in Canada and 197 in North America, Mexico included. Some overlap geographic boundaries. Of those counted for North America, 130 reside in Mexico, 87 in the United States. Of the 87, only 40% are in the high grazing areas of the country. Only 3 species range across the Southwest.

Both farm trade practices and maximum grazing traffic since the Poison Control Centers were established in 1949 demanded an answer. In all probability, pesticides in the pasture and a healthy crop of dung beetles suggest an oxymoron, yet the possibility of manure being handled by dung beetles in a healthy pasture eco-system invited a national attention that agriculture never got. The token laboratory and its work became a casualty to a budget cut, as explained earlier.

It is next to impossible to estimate the number of cattle on farms and pastures in any one year, but the number has to be in excess of 110 million cows, heifers, steers and bulls. Any way you count it, a guess would be 10 cow patties each day, making it possible to compute the overload for a single pasture or farm. College Station research revealed that 82-88% of artificially deposited 5 pound cow pats were buried within a week during a month of peak dung beetle activity, *Onthophagus gazella* presiding. By way of comparison, 96-100% of feces deposited in March through June, with only native beetles present, remained on the pasture surface after one month. Of feces deposited during winter months, 75-95% remained on the pastures after 9 months, 0-40% after 16 months.

The lesson is clear. Losses damage, if not stagger, those trying to reach the bottom line while still using toxicity. Otherwise, propagation and colonization of the right species remain unfinished business for the Department of Agriculture.

One of the researchers mentioned earlier concluded that even one cow pat damages an area 1 meter square, a little more than a square yard. On dry feed the damage is 0.8 meter square, 1.2 meters on silage. A meter is 39.36995 inches. These numbers are the most conservative of many studies. Hypothetical premises are just that, hypothetical. Just the same, give-or-take numbers should suffice for the points being entertained here. That national cow herd mentioned above, give or take, should cover 12,288 acres of pasture per day, or 4,485,295 acres per year.

Give or take tells us to subtract when grazing stops, and it tells us to make computations that differentiate between cows, steers, calves and bulls, defecations moving from 6 for a calf to 10 for a cow. Multiply the grand total times 365, and the total of acres involved that are out of commission rises up to haunt and horrify. Some 13 million cattle are always in feedlots and do not graze. Assuming that 75% of cattle dung is still dropped on pastures, half of which remains on the surface, then it appears, according to Truman Fincher, that 829,474 acres are continuously covered by cattle dung each year.

Undigested dung on grass is at least as damaging as noxious weeds. Cattle will not graze on rank growth, nor will they eat contaminated grass. Cattle refusal of contaminated grass is often 80-83% refusal, this when there is only 16-26% refusal of the same grass with dung being absent. It does not take a rocket scientist to understand why animals recognize the offensive properties of dung. Foremost, it imparts an undesirable taste to the forage. If an animal can satisfy its needs in any other way, it will keep a wide distance between itself and manure still drying in the sun.

Research workers have wide differences of opinion about the range of a cow pat's contamination. As mentioned earlier, figures used by Fincher's College Station research always tend to be conservative. Using the computation which assumes that some 829,474 acres are covered by dung each year on a continuous basis, Fincher computed that 4,147,369 acres of grazing are constantly off duty to grazing because of this contamination.

Guesstimates are guesstimates, no more. Still, the cost has to be astronomical. There are calculations that translate such losses into beef on the hoof, but the point here is pasture loss, and the economic losses can be settled farm by farm with a small calculator. Prices and each passing year make calculations suspect each day. The dairy farmer can construct his own numbers and measure his own profit leakage, this assuming that the dairy cows are maintained chiefly on grass and not fed high energy rations, and that they escape partial or total confinement.

At least 80% of the nitrogen contained in cow manure evaporates, lost to ground cover. The literature is full of citations and footnotes by workers who have tackled the subject. This loss can best be mitigated by the presence of dung beetles. Herds stomping these nutrients into the soil before moving on may help some, but dung beetles when colonized represent Nature's finest development for the purpose of nutrient retention. One researcher duly noted that most of the nitrogen is present in a component of undigested protein which is denatured by bacteria. This is lost by utilization as ammonia. Buried by dung beetles, that same protein envelope has a beneficial effect on grass growth.

Using the factor of 10 eliminations, approximately 50 pounds, the payback has to be measured in tons per day, whatever the national animal count that year might be. The total per year has to be in the neighborhood of 436-438 million tons. At 80% moisture and 2% nitrogen on a dry-weight basis, a single year of Fincher's calculations means the loss of nearly a million tons of nitrogen, representing the economic loss of N.

Failure to support the facilities needed to repopulate the nation's pastures with dung beetles reveals a shortsightedness in public policy that goes beyond criticism. The cost of replacing such a nitrogen deficit runs into the millions, even when computed at yesterday's prices.

There's more. Dung beetles are a valid control for gastro-intestinal parasites, which, like the poor, are always with us, especially in warm climates. Parasites mean morbidity and veterinary costs, medicines included. Parasites do bow to sanitation, however, and this means that they cave in when a good clean-up crew is on duty early. The near wipe-out of these natural janitors in the pasture due to changed technologies over the last half century or more assigns cause to effect. It may be that the range-fed animal has less of a parasite problem than the resident of a tame pasture or the animal that runs low on feed in a rotational system. As feedlot animals readily illustrate, crowding also augments the transmission of parasites. Any concentration of host and parasite in a small area holds in escrow the danger of infestation. Control with pharmaceutical chemistry does not fall under the purview of this narrative. A new look at the efficacy of dung beetles does.

When eggs are passed out of the living host and then allowed to hatch in undisturbed dung, the parasite takeover is off and running. Moisture and temperature preserve infection. Migration of the infection to grass becomes a given. The grazing animal attempts to avoid the infestation, but the life cycle of the parasite completes itself unless interdicted.

This approximates the dung beetle story. Researcher Truman Fincher, now Farmer Fincher, will shake his head in abject disbelief over the monument to the stupidity of man that cancellation of the dung beetle has become. Valued research tells of a nine-fold decrease in infections when pastures were given new residents, dung beetles. Losses invite the same calculation already expressed for nitrogen, grazing areas, etc. Mortality due to parasites is generally reckoned at 0.25%, but mere debilitation is not easily measured. All such arithmetic only hints, yet staggers us, but it never makes the *Statistical Abstract of the United States.*

This much stated, how can it be that manure is a most valued product for the diversified farm, even for pasture? A codicil of sorts may belong here because the quality of manure affects grass, pasture and the paddock environment. Mark McAfee of Fresno, California is a dairyman. He brings his milking parlor into the field, and preserves the health of his animals with sound husbandry. "The bottom line," notes McAfee, "is that once you stop the physiological stress, you stop the production of pathogens."

This comment seems self-evident, taking factory herds into consideration. There, great inventories of antibiotics are used, as well as hormones. These cattle populations are identified with poor manure management, no natural grass, and absolutely no way for the cow to really rest. Because of physiological destruction, factory cows slough off pathogens in manure constantly.

Catherine Berg of the University of California at Davis is a veterinarian. Her tests revealed that grass-fed cows at McAfee's Organic Pastures Dairy dropped manure that contained absolutely no Salmonella, this finding from manure directly evacuated by cows. That is exactly the opposite of results for the infected factory herds that are forever bedeviling Congress for bailout money. McAfee added, "We took our milk to labs here in

Fresno. They added 10 million counts of pathogenic bacteria to each liter of milk. They couldn't live."

Some of the above comments, especially about indifference, may seem odd. Odder still is the fact that nearly everyone who hears the dung beetle story exhibits profound interest. Perhaps this effort under your eyes will cause someone yet unknown to rise to the top and recapture the values. That's what the process of civilization is all about.

Phanaeus difformis *is a ubiquitous native of the U.S. South. It has a very shiny metallic green bronze texture on its shell and forequarters. Its horn is as tell-tale as the cranial appendages of a trail-driven Longhorn steer.*

20

A Codicil of Sorts

Thus we return to the opening line of the prologue. Like John Wayne crossing the desert, the Sun God Ra symbolizes, and is, the Great Scarab. It rolls the fiery orb like a gigantic dung ball across the sky. The earth was the opaque reality, stationary, with night and day governed by the grace of Ra. Burial and resurrection were both achieved by the male alone out of inanimate matter. Sex, of course, was set aside in Egyptian mythology. The scarabs of that ancient land were believed to be male, albeit not the issue of two sexes after union. All males deposited their seed in the legendary dung ball to issue subsequent generations of dung beetles. It has been speculated by Egyptologists that the "pregnant" dung ball in its secure chamber translated itself into the practice of mummifying human remains for resurrection. *The Book of the Dead* is chock-full of scarab symbolism. Egyptologists further speculate that dung burial and dung beetles rising from the soil found cultural ratification in mummy art, in the deep-earth entombment chambers found by Carter, and in world fascination with artifacts reclaimed from the tombs of Tutankhamen and other pharoahs. In the fullness of time, the scarab made the transition from dung crawler to treasured talisman for the necks of stunningly beautiful women.

So far there have been no dung beetles placed on the endangered species list, although certain genera are perhaps the most endangered of all beetles. The toxic rain that was blessed by the establishment of Poison Control Centers in 1949 elevated toxic farming above environmental and biological concerns. The half dozen or more beetle genera considered endangered are losing out not because of toxic farming, but because collectors treasure them for their size, rarity and unique colors. Even entomologists, when speaking of beetle extinctions, point their finger at collectors and merchants. Collectors are much like stamp collectors, forever glorying in a specialty, but they're rarely cognizant of the role these little animals play in the ecology, much less a cowman's bottom line.

It is significant that beetles of many types come under the Endangered Species Act of 1973, which protects — if that is possible — invertebrates in the United States. "Protect a bug?" intoned the masters of the Golden Fleece Award, one senator's concept of Nature's insignificance. Yes, protect a bug could well be our mantra when we realize the various roles of Nature's guardians and destroying angels. Our composters bring microorganisms to the biomass. The dung beetles take the manure to Nature's smallest unpaid workers. By the end of century 20, nine species were named "protected" by the ESA. The American burying beetle leads with its descriptive scientific name, *Nicrophorus americanus.* The rest of the roster marches through the genera of beetledom with not a dung beetle in the lot. Destruction of habitat is supposedly strictly against the law, as is dead animal traffic. Yet new roads and buildings seem exempt. The Fish and Wildlife Service is the agency charged with enforcing that law.

When the curtain came down on the career of the foremost proponent of field colonization of dung beetles, it didn't close down the worldwide interest in, or study of, the creatures. Even as we speak (so to write), new papers arrive, most of them spawned in the manner that Truman Fincher described in the penultimate chapter of this book. Truman in fact came to the San Antonio *Acres U.S.A.* meeting on August 28, 1997 to restate the case for dung beetles and to explain how come it was blood on the moon for a program with so much promise. He had been working with dung beetles for 31 years, and he admitted that establishment of a suitable beetle population in the various parts of the country would be a task on a par with putting a man on the moon, albeit more valuable. Unlike too many in farm research, Fincher grew up on a farm in southwest Georgia. Like many farm children, he was fascinated by those industrious tumblebugs, but he hardly expected to spend a working lifetime studying insects that eat cow dung. He wrote that obligatory Ph.D. dissertation, then went to work for the USDA at the Agricultural Research Service in Tifton, Georgia. At Tifton, he demolished many myths, installed new findings, not the least of which was the lure of swine feces for dung beetles as trap bait, and calculations that defined the profit potential of dung beetle populations to cowmen who had never even considered such a proposition. Some ten years later, he landed at College Station to help service that late-lamented USDA dung beetle program. Now he was bowing out, first explaining to stockmen not how the hogs ate the chickens, but how the beetles made the pasture dung disappear. The same lecture to academia required further explanation and elaboration, he said, because too many experts who farmed behind a desk with a pencil did not comprehend the simple vocabulary of food producers.

Here is the message he left on his answering machine on his last day: "Hello, this is Truman Fincher, a former employee of the USDA's Agricultural Research Service. The ARS, as it is commonly known, is the research organization of the U.S. Department of Agriculture. I am not in my office today. I have been forced into early retirement because ignorant administrators in the ARS decided that I do not need to work on dung beetles any more, even though I have received an average of one request a week from people throughout the country for the last three years seeking information on dung beetles and where to obtain them. ARS administrators have turned a deaf ear to their requests.

"ARS used to be a great organization that did research to help the agricultural community, but during the last 10-12 years, line managers have been replaced by too many people who were born and raised in cities. Many of these managers have never been on a farm or ranch and are out of touch with the kind of research that farmers and ranchers need. In fact, these managers do not want input from farmers and ranchers to help make decisions on what research problems to work on. Some ARS scientists are spending your tax dollars conducting useless research that only benefit themselves. In other words, the ARS of the USDA is being destroyed from within by ignorant managers and the decisions they make."

He explained, as the paragraphs of these pages have endeavored to clarify, that all vertebrate dung lures dung beetles. The Cadillac of dung, however, has been and remains cow manure. Fincher talked about the dwellers, those small bugs that spend their entire lives in the cow's recently evacuated dung. Everyone present knew about the scarab, the tumblebug, which for many of us in the Depression era was our movie, TV and stage play rolled into one. He even hinted at Y-sex, meaning the Y-chromosome that governs gender, ergo work capacity, everything emanating from a dung ball hidden away by the reverse-geared tumbler, or the even more refined brood chamber of the tunneler, Nature's finest composter. These janitors of the landscape usually remain unseen, these architects of the tunnel, and, not unlike the out-of-sight worker in a coal mine or on a remote ranch, they do the most work.

Truman didn't sound like a man who'd been down-sized out of work. If nothing else, he was a spokesman, and his written reports would continue to guide any and all who were willing to see. It was beneath his dignity to write papers without substance just for the purpose of self-glorification. Others might run data factories, but his interest was in the field, which had first claim on his time. Unfortunately, administrators whom Truman characterized as ignorant bought into the chemical-remedy scenario, or perhaps it was the chemical-remedy scenario that bought into them. In any case, their emphasis was on "economy," an expression euphemistically used for sundry considerations, not excluding a repugnance for work. This comment about ignorance is no hyberbole. There is the cited case of the entomologist who concluded a long conversation with the telling question, "By the way, what is a Hereford heifer?"

Fincher could walk an audience through the known knowledge about dung beetles without a note or teleprompter. He rarely referred to the Latinate names of those unpaid industrial "cleaner-uppers." It was the removal of dung that mattered, the calculable value of dung compost once it was delivered to the microorganisms in the soil.

In World War II, the length of a machine gun ammo belt on a B-29 was 27 feet, or 9 yards. "The whole nine yards" has been a going expression ever since. Truman delivered the whole nine yards — everything contained herein — without once resorting to academic obfuscation or talking down to the cowmen and farmers assembled. After all, merely removing dung was not the only worthy goal for Nature's decomposers. They sequestered parasite larvae and made it impossible for these destroyers to resurface and do their mischief. The parasite never picked up by the grazing animal didn't debilitate and didn't subtract from that sought-after bottom line.

The Jeremy Rifkin types want the bovine removed from the planet. After all, it passes gas, and, wonder of wonders, it excretes feces, usually delivering 9-10 payloads a day. Furthermore, it is reckoned that up to half of the acres in the U.S.A. cannot be used by agriculture for anything except cattle or other livestock. Admittedly, not all droppings are on pastures. There are feedlots, dairies and now commercial composters in almost every area trying to cope with the overload.

It is the failure of the dung beetle to increase and multiply in harmony with the increase in the cattle population that most concerns Fincher. That failure is now compounded by bureau inertia and a retreat to intellectual puberty.

Although Truman himself made no such claim, it became obvious, to all who wished to see it happen, that the "greatest invention" was forming up and becoming available to any entrepreneur who was content to stand on the shoulders of giants while completing the task sometime during century 21.

In covering that proverbial picture worth a thousand words, and those thousand words with one demonstration, Fincher issued a requiem for the Argentine Pasos Dung Beetle Project 1969-87. Foreign exportation and importation allow man to give an assist to Nature's interminably slow process, but now those tax dollars were down the tube, and all that the bureau people had to show for it was the epitaph in Gilbert and Sullivan's *Princess Ida:*

Man is coarse and man is plain,
Man is more or less insane.
Man's a ribald, man's a rake,
Man is nature's sole mistake.

Fincher has an even better memorial for his effort. The 15 foreign species imported required exploration, propagation and release, and they remain a living monument and an assist in broad-spectrum recognition of the Scarabaeidae.

The above was a tough act to follow, and I was next on the roster. What more was there to say? It was this business of an agriculture based on the harshest of salt fertilizers and toxic rescue chemistry that came to mind because chemicals can annihilate dung beetles faster than they can be colonized. It is the nature of toxicity that farmers understand even less than they understand dung beetles.

As noted earlier, I have made scant use of beetle nomenclature, and have barely cited the hundreds, even thousands of workers who assembled dung beetle knowledge even before Pliny the Elder and Carolus Linnaeus. Most of these workers added their insight and their volumes to the fount of knowledge after the early part of century 20, but chiefly after 1970. To have cited all of the above would have required the patience of an airline traveler and an indifference to pain. Here is my point. It is the custom to cite the investigator who first described a certain dung beetle. Communication becomes intolerable, if not impossible, when the farmer or rancher must be presumed unfamiliar with much of the grammar of the subject. The Ph.D. will understand names under the heading Hard Consumers: Aegialiidae, Aegialiinae, Eremazinae, Allidiostomidae, Belohinidae, Ceratocanthidae, Chironidae, Diphyllostomidae, Geotrupidae, Athyreinae, Bolboceratinae, Geotrupinae (incl. Taurocerastinae), Lethrinae, etc.

Then there are the Intermediate Consumers and the Soft Consumers. The list goes on as far as the eye can see without materially helping the cowman understand how a small bug presides over his income statement (previously known as a profit and loss statement).

Then there is the matter of a letter that surfaced during this assembling of dung beetle information. It was addressed to a USDA functionary and carried the imprimatur of a representative of the chemical industry. Science, it said, has poisons that will kill the parasites that are drawn like a magnet to a cow dropping. No need to fool around with dung beetles.

That this missive missed the point by a country mile is self-evident to the discerning reader. What about the dung and its value?

I have restrained my commentary on a particular herbicide available under some two dozen brand names, including the now well-known Roundup. The going billingsgate has it that it is dangerous only when the lungs get a fix. In the soil, glyphosate is reputed to feed microorganisms. Since it contains no chlorine, its effect on the atmosphere is supposedly nil. In all such matters, a journalist contacts an expert with credentials and standing, in this case geneticist Joe Cumming, Ph.D., London, Ontario. There are formulations and there are formulations, and many contain some very nasty things. The original Roundup contained a wetting agent better known by the acronym POA. The problem with wetting agents is that they are sheltered as trade secrets.

"Glyphosate itself is quite nasty," said Professor Cumming, "nasty to amphibians — frogs and toads, for instance. This is happening on a grand scale. As for dung beetles, they haven't a chance. There is some evidence that glyphosate does damage the liver, and this clinical damage pertains to human beings. Evidence from South America, a more transparent society, has been stacking up for years. It acts like estrogen. It feminizes and reduces the fertility of man," the professor explained.

The route of mischief is through the skin or via the diet. As more and more Roundup Ready crops come on the market, the regulatory agencies simply raise the tolerance levels for glyphosate in foods. In the original genetically modified crops, an additional gene was added to make the plant insensitive to glyphosate. This never really worked very well, so in order to increase the plant's tolerance, additional copies of the tolerance gene were added. Thus, more and more glyphosate is pumped into food. It was bald-faced collusion with regulators. Farm animals and human beings fed such fare now get a dose that has to have a negative impact on health.

This toxic agent is now fed to crops in every way possible, irrigation by spray, by drip, by flood. Other chemicals are now added to glyphosate to make the solution denser, acrylamide, for instance, which is a plastic and a known neurotoxin. In the soil, it supposedly prevents leaching, but dung beetles can't survive the stuff. As this preparation inserts nitrogen into the soil, it also bequeaths nitrogen to the ambient air. High levels of nitrates in drinking water are also a consequence. The carcinogen connection can be imagined, as can Infant Death Syndrome. Hundreds of chemicals are now in daily use in farm country U.S.A. In fact, the *Farm Chemical Handbook* describes thousands. A comment on only one merely hints at the complexity of the subject and its threat.

The discerning reader will have noted that we have yet to mention the deficit side of dung beetles. Here is Truman Fincher's list of injurious dung beetle species, all of which belong to the Aphodius genus. The larvae of *Aphodius granarius* damages sprouting corn, as happens in Minnesota. It has also been implicated in attacking grass roots in Oregon. Other Aphodius species are called *A. pardalis, howitti, fimetarius, contaminatus, subterraneus, distinctus, pseudotasmaniae,* and *hamatus. Aphodius fimetarius* and *subterraneus* are adults when they do their damage to mushrooms, but all the rest are larval mischief-makers. *Pardalis* like golf turf, as do *contaminatus. Howitti* and *pseudotasmaniae* prefer pasture grass, and *fimetarius* larvae like potatoes. *Distinctus* goes for mint.

The professional bugman will find a thousand omissions, nuances and idiosyncrasies that surely ask for a great enabling transport into the literature. The same is true for helminths. Just the same, I hope this presentation gives the cowman a nodding acquaintance with the subject, and the gardener a new appreciation of the tumblebugs that so entertained us in our youth.

Fincher's career has not been without honors. Don L. Bull, Research Leader, nominated G. Truman Fincher for the 1994 Mallinckrodt Lifetime Achievement Award in Livestock Entomology. He won that award, of course, but long before that he had achieved an international reputation for biological control of biting flies. He accounted for at least 70 scientific publications and has presented dozens of papers to professional assemblies. One citation, again by Don L. Bull, has captured the essence of "the dung beetle man" more than most.

"In recent years, Dr. Fincher has focused his work specifically on predatory beetles that attack immature stages (egg, larva, pupa) of the horn fly. He recognized that the action of predatory beetles would complement those of dung-burying beetles to provide greater negative impact on the horn fly. Dr. Fincher has successfully imported, reared in the laboratory, determined life history and ecological requirements, and released into the field several predator species. This work can logically be predicted to result in the establishment in nature of exotic species of horn fly predator beetles that will work synergistically with dung-burying beetle species previously established in nature as a result of Dr. Fincher's earlier work.

"Dr. Fincher has dedicated his entire professional life to work aimed at the development of sustainable and ecologically sound management of destructive livestock parasites. His successes have been significant. He is respected world-wide as a major contributor to and spokesman for this important area of work; he is a valuable resource for advice and counsel; he conducts his work openly and with a willingness to work cooperatively with those outside his own lab to facilitate and accelerate advances in the successful management of livestock parasites."

Necrophilus hydrophiloides: *the genus and species of the Agyrtidae family mentioned by Mark Sturges is shown here as a mature beetle. Though not a scarab, it seems to be a switch-hitter between dead organic matter, carrion and dung.*

Afterword

As I closed down this report on dung beetles, I had a chance to visit with Kathy DeBoer, an American treasure. From Maple Ridge, Minnesota, she has sent out a message that is certain to have an impact on our drive to install dung beetles on American pastures.

Kathy's garden-waste compost pile has become a nursery for beetles. Without advice or consent, or even much knowledge of Nature's sanitation workers, that epitome of rescue, the St. Bernard dog, both made and saved the day. It was the big dog's dung added to the compost pile that lured several species of small beetles that loved high-protein dung the way Truman Fincher's trapped beetles loved swine manure.

Kathy told me she'd moved quite recently to a facility destined to feature a charter school dedicated to Jane Goodall.[1] "At our old place we had a dog kennel. We had gravel on the runs and a half-acre fence for dogs." The St. Bernard can weigh up to 200 pounds. Its dung output is best described as copious. To save the grass, Kathy scooped up the dung and transferred it to her compost pile.

"The dog dung was sitting on the ground overnight," she recalls, "and by morning it was crawling with Geotrupes." In short order, Kathy DeBoer's compost pile became dung beetle Heaven. "You could scoop up a shovel full of beetles, give them some dung for the road and take them to populate pastures with suitable stock density."

"Food, herbs, and pasture" has become a credo for the lady with the St. Bernards — and dung beetles. She came by her interest in the great dogs when a disabled child needed a cart-puller. Children grow up, but the love of the big dog never grows old. Legends may have helped. The dog with a brandy cask slung under its neck is a fiction of course, the invention of an artist. But the rescue quality of the animal is not.

Historians tell us that the St. Bernard goes back to the mastiff war-dogs of the Egyptians. The genes of herding dogs also seem to figure in the makeup of the animal, which was a farm dog before the monasteries led them to fame. Folklore has it that the St. Bernard would pull a cart from the hills down to a village, house to house. Each patron took an allotment of milk and cheese, and placed the appropriate currency in the cart — all this without supervision.

The most famous dog rescued 42 people during its life. In fact, the dogs usually went out in groups. When they found a stranded traveler, one would stay behind to supply body heat to the sufferer while the others returned to the monastery to fetch monks for the rescue. The rescue of dung beetles via the St. Bernard caught the attention of veterinarian Will Winter.

"Kathy rescues St. Bernard dogs that people discard or abuse. I rescue dung beetles from her compost pile," Winter confirmed. "We have to propagate our own and distribute our own. We'll have to repopulate the grasslands using a few simple tools, a shovel, a bucket, rubber gloves and an environment in which a dung beetle can do its job." One of the species

[1] Jane Goodall lived alone with chimpanzees. In the 1960s and 70s, *National Geographic* got involved. This was well before Diane Fossey became a household name because of movie and media exposure. Jane Goodall is now in her 70s, a constant lecturer, and a hero to Kathy DeBoer.

that likes Minnesota and the northern tier of states is a little brown fellow named *Aphodius granarius.*

"It's the loss of grass that pains," muses Will, a dung beetle aficionado. Winter is associated with Thousand Hills Ranch, a grass-fed beef operation in Cannon Falls, Minnesota. He takes pardonable pride in having colonized beetles on some 30 farms during recent years and hopes to continue that performance each year. When I visited with Winter, he was deep in thought on how to make a dung beetle the insignia for the Thousand Hills enterprise.

"Because farmers have to get back to grass," emphasizes the livestock consultant and veterinarian. "When we use heavy livestock density with a moving mob type of grazing, there's an enormous amount of manure. The first thing I look for on a ranch is evidence of dung beetle presence. If there are no dung beetles, you can make book on present or recent use of wormers." Winter runs the inventory of considerations by the willing listener. "If the minerals are low, if iodine and selenium are missing, we invite parasites." For these reasons and for reasons already stated, Winter doesn't assume worms. "Get a flotation solution, check the manure with an inexpensive microscope, and rise with surprise when a stranger shows up." Dung beetles seem to appear if the conditions are right. Removing dung beetles from their habitat for identification is simple in the extreme. Simply insert the shovel with beetles in a bucket of water. The beetles will float to the top. They can be skimmed off, packaged in a non-plastic container and handed over to a university or Extension entomologist for correct identification.

DeBoer tells of a buffalo rancher who literally saved his herd with dung beetles. The buffalo grower had very poor fertility in his pastures. Will Winter went out and worked with him. Part of the remedy was the introduction of colonized dung beetles. "The fertility went way up," she says.

The story sketched above provides no more than a charcoal image of a movement afloat with hardly a university being aware of it. As farmers and ranchers turn to grass farming, they can expect Omega 3 and Omega 6 on a 1:1 ratio. The loss of pasture cries out for a remedy. Getting rid of bad science becomes a first order of business even for cattlemen hooked on wormers, toxic sprays and strange alchemy. The wormers are easy to dismiss. Diatomaceous earth in feed does a perfect job and delivers no apparent harm to the dung beetle population.

Kathy DeBoer turned to propagating several species quite serendipitously when she shoveled St. Bernard dung into her garden compost pile.

The classic reasons for dung beetle culture are validated each hour of the day.

Will Winter admits that his beetles didn't arrive just because the manure was there. In fact, he installed a sort of dating service on the Internet. Friends share, and Winter accepted a fair share of dung beetles via shovel and pail, with dung on top for the period of migration in the bed of a pickup. Will transports beetles from border to border, taking care to instruct ranchers to rubber-glove transplant the new workers into a fresh cow pie. Nature does the rest, always complying with fecundity and balance. New beetles for 50-100 pies will have the whole pasture superbly populated in record time. Will keeps a log of such farmers. Most Texas dung beetles will not survive in Minnesota, and Minnesota beetles can't make it very well in Arizona, although Betsy Ross of Texas hill country tells me that some of the species indigenous to the north country have also materialized in pasture further south.

All state Departments of Agriculture seem to act as though state borders could keep beetles in or out, for which reason awesome laws, rules and regulations have been constructed. Minnesota can serve as a metaphor for all cold weather states. Its students of entomology study beetles of all kinds, and professionals have inventoried more species with fine sounding names than most farmers can remember. Even so, only a few progressive farmers have taken note of the dung beetle's role, almost all of them still tongue-tied when confronted with "Order: Coleoptera and Family: Scarabaeidae," terms that have already peppered this story.

Some scarab beetles are routinely swept up in traps, notably Canthon and Aphodius. Six species of *Trox*, namely *punctatus, aequalis, unistriatus, tuberculatus, foveicollis,* and *hamatus* have all been caught in carrion-baited traps. *Canthon nigricornis* like deer pellets, *Copris tullius* with its horn, and so the catalog goes on. If passages such as this one whet the appetite of a youngster to become a coleopterist, then these words have not been in vain, incomplete as this report may be. All 50 states, though Hawaii has no endemic species, have their share of the world's 7,000 dung beetle species, each roster overlapping.

The reader will note that the U.S. has never constructed comprehensive dung beetle maps to guide the enterprising grower. Such a map, no doubt, would have to follow the outline of a plant zone map — zone 5, zone 6, zone 4, and so on. Beetle species, with some exceptions, belong with the appropriate plant zone. Withal, dung beetles range far and wide in their zones. Their flying performance may resemble a truck with one or two cylinders missing, and that would seem to cancel out such distant migration.

"I'm not so sure about that," said Betsy Ross of Granger, Texas. Genus and species have shown up in her prize pasture, which is almost a symbol for the ranchers she serves as a leader in the Holistic Resources Management movement. In a territory that Allan Savory of HRM has called fragile, Betsy exhibits grassland swards the envy of any rainbelt grazier. Her best kept secret is compost extract, which she makes as an art, and which HSM ranchers treasure. She also has beetles, one of which has been identified as a member of the Agyrtidae family.

"Get rid of the chemicals," she counsels, "and present an environment with food. They'll come," she assures all who pause to listen. Listeners are those who attend her course work or simply visit the ranch. "You can give the beetles an assist, certainly, but they'll come on their own." Yes, if they have something to come to that doesn't exterminate them. Yes, despite their clumsy form of mobility, they do get around. They ignore political borders with a disdain usually reserved for a bad stage act.

"You can't use Ivermectin and expect to have a healthy pasture," Winter affirms. In a lush pasture, it's the tunnelers and dwellers that preside. With the grass-fed concept expanding exponentially, and consumers rejecting science-without-humanity, a rollover in farm technology can be expected within a generation, GMOs, terminator seeds, and Frankenfoods notwithstanding. Ivermectin is one of the nemeses of the dung beetle. As ranchers move over to diatomaceous earth or even Cydectin (a Fort Dodge product) as wormers, dung beetles are given permission for life.

According to Betsy Ross, dung beetles have proliferated throughout Texas to the degree that more and more ranchers are abandoning beetle-killing pharmaceuticals. Betsy's parting shot as I closed down the manuscript was simply, "All of Texas's sustainable-growth clients have dung beetles."

Soil biology programs also help. In recent years, Elaine Ingham, the ex-Corvallis, Washington Ph.D. who counsels compost tea, also has become more than a subliminal voice in making the good earth dung beetle welcome. Here and there, enterprising young people have come up with improvised packages that serve as starters.

One nationally known figure who has picked up the gauntlet of challenge is Mark Sturges of Bandon, Oregon. Mark is a little giant in the image of former House Speaker Carl Albert. Under his black Stetson is a great head of hair and the kind of genius that Einstein spoke of when he lauded naturalists who sought to understand the plan of Nature, chance, or God. Sturges has a starter package that seems to tolerate climate well beyond the crop zone range.

The road to Damascus has many branches. Mark found his at a Healthy Soil Symposium in Cottage Grove, Oregon. The speakers were Elaine Ingham and Andy Moldenke, the latter an entomologist with credentials and standing. During a break, Mark told Ingham about worm tea leachate from his compost bins. He reported results with plants with such enthusiasm that it prompted Elaine to make the scene in person. Ingham's calling card is compost tea. It was this technology that kicked open the door for Mark when he made his superb compost "using manure from all my creatures." He harvested llama and alpaca dung with the eager eyed ardency of a Death Valley Scotty hunting for gold. "Beetles just started to show up," Mark now relates. "They were first observed as larvae." Andy Moldenke announced the presence of a brand new family.

The beetles that Mark found in his compost bin are not dung beetles of the Scarabaeidae family. The family name is Agyrtidae, and its common name is "primitive carrion beetle." Its first name (genus) is *Necrophilus,* species *hydrophiloides,* suggesting a love of water. Al Newton of Chicago's Field Museum of Natural History made the identification in this case, for which reason he also accounted for recent classifications and bestowed the highly scientific moniker along with its history and discovery thus: "*Necrophilus hydrophiloides,* loosely and liberally translated means 'lover of the dead, lover of water'."

Larva of the Agyrtidae family do most of the eating, dung or otherwise. The mature adults seem to have one principal occupation, reproduction.

Aphodius fematarius is a red beetle, not a tunneler, yet a consumer of a prodigious amount of manure. "It probably classifies as a dweller," Mark

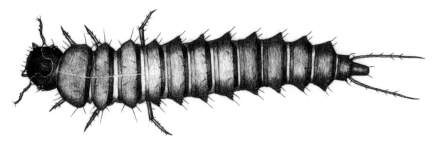

Agyrtidae larva

allows, if the sole diet has to be dung. This one is the size of a big ladybug without spots. You'll always find it living on llama and alpaca dung. This is not to say that Aphodius shuns cattle dung. Mark's locus is in the proximity of a flood plain. Often, dairy cows are moved to higher ground. Manure collects. So Mark sent about 2.5 gallons of raw manure to Pat Richardson, a professor at Texas University in Austin. She reported back that the organic dung manure contained over 90 species of all kinds of beetles.

Mark came by his knowledge quite naturally. He had a compost bin that reached back to the days when he grew countless varieties of chili beans. These garden leavings, coupled with alpaca and llama dung invited in beetles, dung beetles, water scavengers, hister beetles, rove beetles and beetles of the rare Agyrtidae family, a manure and pumpkin pulp consumer. North America is said to have six species, although their genera may still be disputed by some experts. "With their friends the gazella, these are the true masters of soil restoration," Mark opines. The same species seemed to show up at Betsy Ross's farm in Texas.

Some of Mark's beetles compost two tons of pumpkin pulp each year. Still, animal dung is the name of the game, and Mark has an enviable reputation in Oregon among those seeking to colonize dung beetles. In turn, many farmers help their neighbors if they have developed the insight required for elimination of poisons. Where do the dung beetles go when the Coquille River floods for months at a time? "I don't know," Mark admits, "but they're back as soon as there's non-toxic dung." This penchant of dung beetles for just showing up was illustrated earlier when it was noted that *Onthophagus taurus* came to Florida out of the blue even before lawmakers started legislating biology.

Dennis Stoltzfoos moved to his Live Oaks, Florida farm some four years before I asked him about his dung beetle population. "There was no activity at first," he revealed. "Then we got the egg mobile up and running, and the chickens gave the dairy cow droppings a workout. About a year after that, change came on, gradual at first, then rapid. Now the dung pats disintegrate."

Dennis figures a beetle is a dung beetle if it consumes and buries dung. Always busy battling a bureaucracy that wants to close down his fresh raw milk dairy, he hasn't had the time to get settled entomological identification, "but there are quite a few species." Once his chickens and cows changed the ecology, "they came," he said. He believes that mineralization attracts dung beetles. Mineralization in the absence of toxic chemistry further improved the quality of the manure which in any case was top quality because of kelp and Azomite and other mineral input. Dung beetles like quality. They perish when their breakfast is laced with the residue of toxic wormers.

Finally, I do not want to leave the reader with the idea that the culture of dung beetles passed away because a program at Texas A&M was cancelled, or because CSIRO ran out of money. Universities still teach students about that useful animal, and some — Australia's University of Queensland, for instance — sometimes morph earlier findings from CSIRO into working programs replete with a roster of objectives suitable for any environment. The recent Queensland Dung Beetle Project sailed forth with the cooperation of the Department of Primary Industries and Fisheries. The goals were education, repopulation, colonization and virtual elimination of toxic genetic chemicals from pastures, with wormers restricted to those that delivered no harm to dung beetles. Of special interest to many American cowmen is the extent to which cattle growers participated in the programmed renaissance. A stated goal was orientation of the extended community on the value of Nature's finest worker.

As I handed over my manuscript to the publisher, I remarked that if a NASA team should discover a bacterium on the moon or on Mars, the scientific community would become delirious. Yet we have overlooked a world inventory of valuable beetles, and ignore a life complex that may or may not exist light years away, but certainly not in places likely to be reached in the life span of a human being.

Our era of a toxic fix for every farm is bound to pass. Bad science and big business await the vengeance they so richly deserve. Always, man proposes and God disposes. Not for dung beetles alone must toxic genetic chemicals be removed from the environment, but for mankind.

The next time you're in Santa Fe, New Mexico, go to the Orehouse Restaurant. Ask the chef or the management for a tour of the insect collection, dung beetles included. It may be the greatest private collection west and east of the Smithsonian. Collecting is often the open sesame to scholarly interest and the profession of the coleopterist.

All the above stated, it is nevertheless a fact that few ranchers really cooperate in handing over a few treasured dung beetles to their neighbors. They may not have the time. But I rather expect that the effort falters for lack of communication. Experts on the subject tell me that print, telephone, the TV screen, and even a personal DVD can be only 7% effective in communication. The most effective form is personal contact with a facial expression that says, "This is the greatest thing that has happened to me today." That's worth 64%. The rest is made up by tone of voice. The fast talker, the brow-beater or the head-cracker cancels out the sound of his own voice. In short, you have to be on site where a farmer has dung beetles. Personal contact with a bucket, a shovel and trunk space is almost always the ticket for closing the deal. The best-case scenario is to trade one species for another for greater dung beetle proliferation.

As these lines are set down, quid pro quo intelligence reveals that insectaries would like to take on the propagation chore, but bureaucracy gets in the way. Each state wants to tell the dung beetles where they can go. Chemical companies hold the hammer over bureaucracies and during the past five or six decades have become the minders of the land grant colleges. And of course a federally funded program is out of the question.

My late associate and mentor, C.J. Fenzau, once gave me a paean to ecology that I'm going to recite here: "When living life emerges into the ultimate objective of our nation, only then can we begin to construct the understanding of the ecology of the soil and its creatures and be able to benefit from its minuteness and wondrous patience. Nature has no limit on time. She is patient and forgiving. She is able to repair herself from the ignoble treatment of man in spite of his tremendous physical capacity for destruction. As we continue to replace Nature, we assuredly prevent the development of our mental capacity to learn and fully complement Nature, a requirement expected from us in permission for life."

The dung beetle awaits an answer.

Appendix 1

A Dung Beetle Demonstration Farm

MATERIALS

Lumber CCA treated lumber or redwood
BOTTOM (2"x4")31" long. Cut a generous groove ⅞"wide by
½" deep down the entire length of the board.
ENDS 2·(2"x2")21" tall. Again, cut a generous ⅞"wide by
½" deep groove the entire length of the board. At the bottom
end of each piece cut the board to leave a ½"deep, ⅞"wide
tongue to fit into the groove in the bottom piece.
BRACES. Make 2 from 2"x4" to glue to ends and
screw to the bottom.
TOP (1"x2")26" long. Cut a generous ⅞"wide groove ¼"
deep through center. In the center of the groove cut a slit
the width of the saw blade, 16" long through the
board for an air slit.

PLEXIGLASS

2 viewing sides 3/16" thick, 25"wide by 21" tall.
2 end spacer strips ½"thick x ½"wide 20.5" tall.
1 bottom spacer strip ½"thick x ½"wide by 25" long.
3 support circles (or □ or △) ½"thick about the
diameter of a quarter to keep sides from bowing.

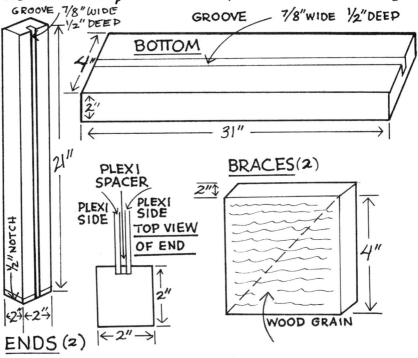

Glue braces to each end.
Secure with screws through bottom.
NOTE: wood grain of brace should be
parallel to bottom board.

BOTTOM

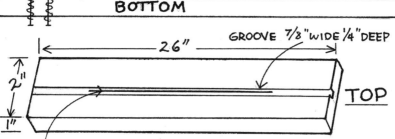

GROOVE 7/8"WIDE 1/4"DEEP

26"

2"

1"

TOP

Saw blade width cut 16" long in center of groove
clear through the board for air vent.

Glue all spacer strips and circles to one plexiglass
side. To space circles: 1 in center 16" from bottom
other 2 go about 8" from either side and 6"from bottom.

Glue and screw all pieces into place.

Notch top 1/2"x1 1/2" on each end to fit over ends.

When all assembled,
drill through
plexiglass side, through
circles and other side.

*DIAGRAMS ARE NOT TO
SCALE! FOR ASSEMBLY
ONLY!*

Secure with nuts and
bolts to keep sides from
bowing in or out.

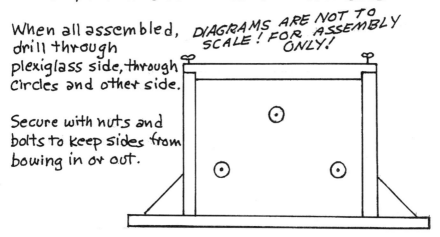

Drill through top into ends to secure with wingnuts.

Appendix 2

A Poem by Mark Sturges

We Must Be Rich

Was there ever
A naturalist who proclaimed,
"One could always
tell the wealth
of a man
by the size
of his pumpkin pile
in late November?"
As the pile grew,
so did the
populations of
forest inhabitants.
Raccoons were pretty
obvious by their
seed-stealing holes
punched in soft spots.
They seem only
interested in the big white seeds.
Composting pumpkin pie
appeared not to
be their forte.

The real connoisseurs
of the large gold squash
are a six-legged
black animal
called an Agyrtid beetle,
the king and queen
of the fall to winter cycle.
Not much is
seen or written
about these animals.
A natural history
pamphlet out of California
reports them being seen
in only five states:
Kentucky, California,
obviously Oregon,
maybe Washington.
I don't know
the other one.
The reference
says they show up
when the weather turns nasty.

Two weeks into
October we had two days
of cold rains.
Magic! There they were!
The first scouts
were in the compost bins.
Swallows to Capistrano,
Agyrtid beetles
showing up at Chili Nervanos.

Last year
they decomposed
one and a half tons
of pumpkins in
two and a half months.
The Agyrtids look
over the pumpkin supply,
and then they breed their way
into proper numbers
to accomplish the job.

If one were
a beetle voyeur,
the good life
would be at hand.
Less than a month
after seeing the
first adults,
larvae are wriggling
everywhere.
Pumpkins are becoming
the richest of soil.
Almost before
one's eyes
tired Jack o'Lanterns
become the bordellos
    and nurseries
of these great
soilmakers.

The pumpkins rest
on tons
of alpaca and llama dung.
The Agyrtid larvae
go back and forth,
sometimes eating
the big orange vegetables;
other times the llama berries
are laced with the slithering
dung-eating larvae.
Pumpkins, Manure, Beetles:
eco-systems that
could save the planet.

It rains.
The winds blow.
It gets colder and colder.
The Agyrtids stay around
producing compost
and more compost
until April comes.
One day they aren't there
    anymore.
Where did they go?
Where did they come from?

Calls start coming in
from all over the country.
"We need that
scarab compost."
They really don't
say that.
Everyone thinks the
earthworms do it all, but
the big pumpkin
pile with its
accompanying totes
of llama and alpaca berries
say it
just isn't true.
In January,
this heap will
be wiggling with
beetle larvae.
It's November.
Our pile is huge.
We must be rich.

— *Mark Sturges*

Acknowledgment

In general, I have tried to hand off as many credits as the flow of the narrative would allow. It is no oversight that I now mention Jake Voegeli, the good right hand anyone with diminished eyesight needs to finalize a manuscript. If all errors are my own, the advanced polish of this story is due to Jake's attention.

My long-time associate, Margaret Merrill, a University of Virginia librarian, turned up more literature than I thought existed. Retrieval systems have changed several times since World War II, in effect often making archival information unavailable to writers. Her ability to rise above this impediment suggests resurrection of hard copies long believed extinguished. She made available long-out-of-print tomes and numerous research articles at the drop of a hat, and is well beyond the amateur or dilettante stage in comprehension of dung beetles herself.

Malcolm Beck, the author of *The Secret Life of Compost,* has added his insight, as have all those interviewed for this book. Will Winter of Minnesota and Mark Sturges of Oregon both deserve mention here because of our telephone bills that look like the national debt, the latter because of his "We Must Be Rich" poem, which lionizes compost and beetles that like pumpkin pulp as well as alpaca, llama, and cattle dung. It appears in his small volume of poems, *The Return of the Fertilizer King and Other Tales,* and is presented as Appendix #2.

Former college chemistry professor, now farmer, Ed Fashing of Sturgeon, Missouri, was ever available as a consultant on chemical matters, some few of which are properly a part of this narrative. Coleopterists at Cornell, Montana State, and Texas A&M shared their insight, as did Jim Nardi of the University of Illinois at Urbana-Champagne. The art work in the Afterword for Agyrtidae, larva and adult, is his, and is a part of his *Life in the Soil, A Guide for Naturalists and Gardeners,* a University of Chicago release. The rest of the art work in this volume was drawn by Skeeter Leard of Fullingim-Isenhour-Leard Galleries in Socorro, New Mexico.

Bryan Kight, the intrepid artist/designer who creates the finished pages of *Acres U.S.A.* magazine each month, converted a literal mountain of paper into this graceful volume you now hold in your hands.

The tireless librarians at the Raytown, Missouri branch of the Mid-Continent Public Library received my multiple, far-reaching interlibrary loan requests with a smile and accepted the challenge to deliver the requested research. Without their help this book would not exist.

Locals and family who never tired of my dung beetle lore also deserve thanks, having endured until I ran down. But it was on-scene observation that piqued my interest in the first place, which hasn't wavered very much since I placed, with childish deviltry, an obstacle in the path of my first tumblebug. They must have understood the game we played and forgiven the gentle teasing delivered through childish ignorance. This much must have been programmed into their DNA.

Index

2,4,5-T, 10
2,4-D, 10
Acres of Diamonds, 8
acrylamide, 185
Aegialia rufescens, 83
Aegialiina rufescens, 121
Aegialiini, 115
Africa, 14, 15
Agricultural Research Service (ARS), 181-182
Agropyron spicatum, 156
Agyrtidae, 52, 192
Albrecht, William A., 46, 47
Altars of Unhewn Stone, 139
Amazonian rain forest, 153
ammonium nitrate, 170
anhydrous ammonia, 170
Animal and Plant Health Inspection Service, 27
Animal Parasite Research Lab, 22
anthelmintics, 154
Anthropoda, 114
aphids, 106
Aphodiinae, 96, 117
Aphodius fematarius, 192
Aphodius granarius, 185, 189

Aphodius porcus, 96
Aphodius rufipes, 56
Aphodius sallei, 120
Aphodius, 96
Aphrodiines scarabs, 82
Aquinas, St. Thomas, xxi
Argentine Pasos, 67
Aristophanes, xx
Aristotle, 115
arthropod parasites, 154
Ascarops strongylina, 125
asteroid, 128
Astragalus mollisimus, xxiii
Atholus coelestis, 130
attraction index, 147
Australia, xviii, 129-130
Australia, agriculture, 31
Australia, beetle species, 121-122
Australia, ecosystem, 172
Australia, importation, 37-38
Australian Museum, 42
Australopithecus boisei, 13
autotechnicon, 124
avermectins, 155
Axtell, R.C., 76
Azomite, 194

baited traps, 147
"Ballad of the Scarab, The," ix-xi
banana, sexless, 106
Barrymore, John, 26
beardless wheatgrass, 156
Beck, Malcolm, 157, 159
bees, 98, 106
beetle, feeding styles, 117
beetle, flyers, 54
beetle, horns, 54
beetle, recipes, 149
beetle, self-defense, 54-55
beetle, sex habits, 101
beetle, sneaks, 101
beetle, stages, 49 53
beetle, testicles, 100, 101
beetles, and climate, 137
beetles, large, 137
beetles, small, 137
Beirne, B.P., 156
Bellamy, Charles L., 157
Berg, Catherine, 177
Bermuda grass, 4
Berra, Yogi, 42
Bertrand, M., 155
beryllium, 128
Birds of Killingworth, The, 67
bison, 11
blister beetle, xx
Blume, Richard, 16, 25, 61, 65
boll weevil, 114
bolus method, 7
Book of the Dead, xix, 180
Bornemissza, George, xiii, 15, 29-30,
    75-76, 97, 172
Bornemissza, George, biography, 129
Bos indicus, 103
Bos taurus, 103
bovine removal, 183
Bram, Ralph, 60
British Columbia, 156
Bromfield, Louis, 46
brood ball, construction, 85-86

brood ball, eggs, 104
brood ball, facts, 84-86
brood balls, xix, 5
brood care, 103
brood, chambers, 86
Bruce, W.M., 66
Bryan, R.P., 76
buffalo, 11
buffalo fly, 40, 66
buffalo rancher, 189
Bull, Don L., 186
burrowers, sexual behavior, 100
Burton, Glenn W., 18, 59
burying beetle, 52-53

C.E. (Common Era), 137
Callahan, Philip, xviii, 10
Canadian Journal of Plant Science, 156
Cantharus, xx
Canthon pilularius, 79, 173
carrion, 104
carrion beetles, 52-53, 192
Carson, Rachel, 2, 72
cattle, parasite injected, 153
cave beetles, 52
cerambycids, 149
Chapman, Jonathan, 6
chemical usage, effects of, 47
chitinous membrane, 52
chromosome, floating, 107
chromosome, y arms, 108
chromosomes, mice, 110
cicadas, 49
class, 114
clay powder, 135-136
cloning, 108
coffin beetle, xxi
cold climate beetles, 115
cold climates, 137
Coleoptera, xxiii, 114
coleopterist, 114
Coleopterists Bulletin, 153
College Station, Texas, 17, 59, 65

colonizing beetles, 39-40, 42
Common Dung Beetles, 39
Common Dung Beetles in Pastures of Southeast Australia, 35
Common Dung Beetles in Pastures of South-Eastern Australia, 123
Commonwealth Scientific and Industrial Research Organization (CSIRO), 32, 35, 36, 122-123
compost tea, 191
Congressional salaries, 3, 168
contaminated grass, 175-176
Conwell, Russell, 8
Copenhagen, Denmark, 156
Copris incertus, 120
Copris minutus, 120
Copris pristinus, 51
coprophagy, 117
Coprophanaeus telamon, 153
corn borers, 67
costs, without beetles, 132-133
Cowboy Arithmetic, 168
cowpat insects, 25
cows, and eating, 22
Crick, Francis, 58
crows, 67
crude protein, 156
Crumbine, M.D., Forest, 11
CSIRO, 32, 35, 36, 122-123
Cumberland Island, 18
Cumming, Ph.D., Joe, 184-185
Curaçao, 158
Curculionidae, 51
Cydectin, 191

dairyman, 177
Dante, xxi
Darwin, Charles, 49
Davis, R., 76
Davis, Walt, 1-9, 138
de Soto, Hernando, 146
DeBoer, Kathy, 187, 189
deworming agents, 7

diatomaceous earth, 5, 189, 191
Dichotomius carolinus, 161
Divine Comedy, xxi
DNA, dung beetle, 108
DNA, mutations, 106
dog dung, 147, 187-188
Double Helix, The, 57
drenches, 171
dung, acreage covered/yr, 175
dung beetle, behavior, 74
dung beetle, benefits, 21, 73, 133, 139
dung beetle, body, 35
dung beetle, captured, table, 148
dung beetle, chromosomes, 74
dung beetle, comparisons, 92
dung beetle, cows, 145
dung beetle, decline, 47
dung beetle, deficits, 185
dung beetle, demonstration, 8
dung beetle, diagram, 34
dung beetle, dollar savings, 163
dung beetle, egg stealer, 153
dung beetle, farm box, 197-198
dung beetle, feces, 90
dung beetle, importance, 75
dung beetle, importation, 26-27
dung beetle, injurious species, 185
dung beetle, lay names, 114
dung beetle, lifecycle diagram, 48
Dung Beetle Mania, 42
dung beetle, mating, 109
dung beetle meeting, 16
dung beetle, needs, 7
dung beetle, pasture populating, 190
dung beetle, pencil, 120
dung beetle, per cow pie, 148
dung beetle, pig, 120
dung beetle, pollution elimination, 171
dung beetle, published papers list, 131
dung beetle, renaming, 120
dung beetle, research, 59-60
dung beetle, results, 62

dung beetle, roster, 119-120
dung beetle, sally, 120
dung beetle, sizes, 73-74, 130
dung beetle, stock rates, 97
dung beetle, threshold, 39
dung beetle, transporting, 190
dung beetle, trapping, 124
dung beetle, traps, 147
dung beetle, U.S. survival, 80-81
dung beetle, uses, 145
"Dung Beetles and Dispersal of Cattle
    Dung," 156
dung beetles, and weeds, 138
dung beetles, Kleberg natives, 122
dung beetles, moisture, 80
dung beetles, parks, 144
dung beetles, phallus, 100
dung, burial, 15, 75
dung, colonizing insects, 154
dung, decay rates, 154-155, 156
dung, role of decomposition, 154
dung, recycling, 163
dung, percent buried, 174-175
dweller, facts, 80-81
dwellers, cold weather, 82
Earth geology, 136-137
*Ecological Studies of Dung Beetles
    Serving as Intermediate Hosts for
    Nematode Parasites of Swine in
    South Georgia,* 125-126
"Effect of Excessive Accumulation of
    Manure on Soil Arthropods in
    Grazing Areas," 155
egg mobile, 193
eggs, gazella, 144
Egypt, xviii-xx
elephant beetle, 14
elephant dung, 15, 71
endangered species, 180
Enlivened Rock Powders, 30
eremazina, 121
estrogen, 185
Euoniticellus fulvus, 123

Euoniticellus intermedius, 15, 26, 29,
    62, 65, 153
Euoniticellus intermedius, details, 119
Evans, Arthur V., 157
Evolution of the Veterinary Art, xx
excrement, frequency, 143
excreta names, 141-142

Fabre, Jean Henri, xiii
face fly, 25
family, 114
Farm Chemical Handbook, 3, 20, 143
farm history, 46
farm revenue, 20
fatty acids, xv, 168, 189
Faulkner, Edward, 46
featherwing beetle, 53
feces, parasitic worms, 76
feed lot, xv
Feehan, John, 32, 38-39, 42
Feehan, John, park beetles, 144
females, 103
Fenzau, C.J., 195
feral pigs, 146
Fincher, George Truman, xiii, xxi, 6, 8,
    17, 21, 30, 145-146, 157, 163, 173
Fincher, George Truman, biography,
    58-61, 65-66
Fincher, George Truman, dissertation,
    70, 75-77
Fincher, George Truman, life over-
    view, 181-183, 186
Fincher, George Truman, obituary,
    61-62
Fincher, George Truman, parasite
    experiment, 153
Fincher, George Truman, phone mes-
    sage, 181-182
Fincher, George Truman, thesis, 125-
    126
firefly, 109
flies, and disease, 20
flies, propagate, 20

Florida, 193
flotation solution, 189
flotation technique, 42
fly costs, 20
fly predators, 130
Foods Index Newsletter, 149
Fort Dodge Laboratories, 33
fox dung, 147
France, dung experiment, 154
Franklin, Rosalind, 57
Frontier Doctor, 11

Gaddes, Alexander, 33
Garrett, Howard, 157
gastro-intestinal parasite, 176-177
Genera Plantarum, 117
genetically modified, 185
genetically modified grain, 149
genus, 114
geological ages, 127-128
geological eras, 14
Georgia islands, 17, 18, 173
Geotrupes, 188
Geotrupes egeriei, 167
Geotrupes spiniger, 122
Geotrupes stercorarius, 96
Geotrupinae, 117
Gibbons, Euell, xi
Gilbert and Sullivan, 183
Gillard, P., 75, 76
global warming, xiv-xv
glyphosate, 184-185
Golden Fleece Award, 32
Gongylonema spp., 126
Granger, Texas, 191
grassland atomic waste, 142
grazing, xiv, 139
grazing management, 6
greenhouse gases, 9, 170
Gromphus binodis, 120
Gromphus lacordairei, 57, 120

Haematobia irritans, 158
Haldane, J.B.S., xiii
Hard Consumers, 184
Harris, Robert, 60
Hasan, Abu, 23
Hawaii, 130
hectares to acres, 23, 100
high-protein feed, 5
Hister caffer, 130
Hister chinensis, 130
Hister nomas, 122, 130
Hister scissifrons, 130
Histeridae, 62, 122, 172
Ho, Mae-Wan, 102
hogs, wild, 146-147
Holistic Resource Management
    (HRM), 6, 191
honey bees, 106
horn fly, 11, 61, 67
horn fly, costs, 24, 158
horn fly, emergence, 157-158
Horus, xix, 49
human excreta, 149
Hydrophilidae, 130, 172

Ice Age, 128
igneus, 124
India, 149
infections decrease, 177
inflation, 2
infrared, xviii, 10
Ingham Ph.D., Elaine, 191-192
Inordinate Fondness for Beetles, An, 157
insect sperm, 110
Insecta, 114
intermediate hosts, 125-126
International Code of Zoological
    Nomenclature, 115
International Journal of Parasitology, 154
ionized chemicals, 72
Isis, xix
isohyet line, 80
Ivermectin, 191

Jackson, Wes, 139
Johnson, Rex, 65

Kadiri, N., 154
kangaroos, 144
kelp, 194
Kerrville, Texas, 65
Khan, Genghis, 31
King, K.L., 155
King Minos, 110
King ranch, 122
kingdom, 117
Kirk, A.A., 156
Kleberg County, Texas, 122
klepto-parasite, facts, 96
Krell, F.T., 153
Kruger Elephant National Park, 15

La Brea tarpits, 51
lactones, 154
Lassiter, Dan, 60
Latin language, 118
Leakey, Louis and Mary, 13
Leard, Skeeter, 120
Liatongus militaris, 62, 119, 127
Lindquist, D.A., 66
Linnaeus, Carolus, 50, 56, 116-117, 118
Linne, Karl von, 116
Lisle, Harvey, 30
livestock emissions, xiv
Longfellow, 67
longhorn beetles, 54, 149
Lumaret, J.P., 154, 155

Macqueen, A., 156
Mallinckrodt Lifetime Achievement Award, 58
manure eliminated, 7
manure pathogens, 177
manure removal, 73
McAfee, Mark, 177
McKinley, William, 45

melolonthine scarabs, 149
Mendel, Gregor, 105
Mendeleyev, 72
Merck, 33, 39
Miami Zoo, 141, 142
Micraegalia pusilla, 121
micro-dwellers, 80
mineralization, 193
Minnesota, 189-190
Minotaur, 110
Moldenke, Andy, 192
Monsanto, 39
moxidectin, 153
Mozart, Wolfgang Amadeus, 118
mummy, 180
mycophagy, 117

naming system, 114-115, 117
Namosella fungi, 53
Nardi, Ph.D., Jim, xvi
Necrophagous americanus, 45
necrophagy, 117
Necrophilus germanicus, 53
Necrophilus hydrophiloides, 187, 192
nematode parasites, 72, 154
neo-XY, 107
nest eggs, 103
nesting, 103, 110-111
neurotoxin, 185
New Orleans 1976, objectives, 65
New Orleans 1976, questions, 64
Newton, Al, 192
Nicrophorus americanus, 180
nitrogen, cowpat, 9
nitrogen loss, 75, 131-132, 169, 176
nitrous oxide (N2O), 9, 170
Nixon, Richard, 2
North American Representatives of the Tribe Aegialiini, 120-121
NOX, 9, 170
nutrient loss, 169
nutrient waste, 169

O'Neill, Tip, 2
ocean solids, 73
Oink Smith, 120
Olduvai, Gorge, 13
Omega 3 and 6, 168, 189
Onitis alexis, 36, 37, 62, 119
Onitis aygulus, 123
Onitis vanderkelleni, 89, 120
Ontherus sulcator, 120
Onthophagus, 114
Onthophagus binodis, 36, 120
Onthophagus bonasus, 95, 119
Onthophagus coerestae, 51
Onthophagus depressus, 113
Onthophagus gazella, xxiii, 6, 7, 13, 21,
    26, 40, 47, 59, 61, 64-65, 97, 145,
    153
Onthophagus gazella, details, 119
Onthophagus nigriventris, 120, 135
Onthophagus nuchicornis, 156
Onthophagus pennsylvanicus, 120
Onthophagus sagittarius, 120, 141
Onthophagus taurus, 1, 8, 16, 21, 62,
    85, 117, 144
Onthophagus taurus, details, 119
opossum dung, 147
Oppenheimer, 168
order, 114
Oregon, 191, 193
Orehouse Restaurant, 194
organic matter, benefits, 164
Origin of Species, 49
Osiris, xix, 49
ovarioles, 104
ovary, 112
Ozona, Texas, 159
ozone, 9

paddock tunnels, 32
parasite larvae, 11
parasite mortality, 177
parasites, xiv
parasitism, 66

parity, xxii
parthenogenesis, 106
Pasiphae, 110
pasture dung, interment, 132
pasture forage, 76, 139
pasture management, 132
Pedobiologia, 154
pelletizing forage, 174
Peloponnesian War, xx
Peranus maindroni, 130
Petit, Bob, 69
Pfizer, 39
Pfrommer, A., 153
Phanaeus igneus, 124, 125
Phanaeus vindex, 69, 72, 124, 125
Phanaeus difformis, 179
pheromones, 111-112
Philonthus flavocinctus, 62, 130
Philonthus minimus, 130
phoretic mites, 76
Phosnet, 5
phosphorus loss, 132, 169
photosynthesis, 73
phylum, 114
Physocephalus sexalatus, 125
pigs, wild, 146
plant seeds, 116
Plowman's Folly, 46
Plusiotis bejera, 55
Plusiotis gloriosa, 55
Podolinsky, Alex, 30
Pollution Control Act, 18
potatoes, sexless, 106
*Potential Value of Dung Beetles in
    Pasture Eco-Systems, The,* 100, 163
Princess Ida, 183
Principles of Botany, 117
Proxmire, William, 32
Pseudocotalpa giulianii, 151
pumpkin composting, 193

Queensland Dung Beetle Project, 194

raccoon dung, 147
radiation, 124
radio-active isotope, 41
red beetle, 192-193
Reinecke, R.K., 76
Requiem Mass, 118
research scientists, 152
Revision of the World Species of the Tribe: Aegialiini, A, 115
Rhone Merieux, Inc., 157
Richardson, Pat, 47
Richardson, Pat and Dick, 63
Rifkin, Jeremy, 183
Ripley, Robert, 92
roller, back parasites, 159
roller, nests, 111
roller, pairs and single, 159
roller, repellent, 86
roller, sexual behavior, 100
roller, sexual reproduction, 86
Roosevelt parity bill, xvii
Roosevelt, Theodore, 45
Ross, Betsy, 190, 191
Roundup, 184, 185
RRR, 69
Rubin/Haas Company, 157

Salmonella, 177-178
salt fertilizer effects, 173
San Antonio, Texas, 181
Santa Fe, New Mexico, 194
Santalus parallelus, 130
saprophagy, 117
Savory, Allan, 6, 12
scarab, chromosomes, 107
Scarab Windrow Turner, 84
Scarabaeidae sacer, xviii
Scarabaeidae, 24, 49, 114, 172
Scarabaeidae, sub-families, 117
Scarabaeinae, 117, 118
scarabesis, 149-150
scarabs, 55
screwworms, 158

secretions, antibiotic, 90
semen, 110
sex locus gene, 107
sex-determining system, 107
sexless reproduction, 106
sheep, Mediterranean, 155
sheep pastures, 155
sheet composting, 163
Sheridan, General, 12
Silent Spring, 2, 72
Silphidae, 52
Simmons, Leigh, 100
Sims, Fletcher, 15, 83-84
Sisyphus rubrus, 99, 120
Sisyphus spinipes, xvii, 122
slime mold, 106
Smith, T.J.R., 156
Smithcors, J.F., xx
soil conservation, 46
Soil, Grass and Cancer, 12
Southwestern Entomologist, 157
species, 114, 115
species count U.S., 130, 147-148
species counts, 174
Species Plantarum, 117
sperm, warehousing, 107
Sphaeridium, 130
Sphaeridium scarabaeoides, 130
spiders, 98
Spratt, D.M., 154
SRY, 109
SS PRP, 69
St. Bernard dog, 187-188
Staphylinidae, 62 172
Staphylinoidea, 52
stebnicka aegialiina, 121
steers, parasitized, 22
Steffen, Bob, 67
stem-cell research, 108
Stewart, T.B., 76
Stoltzfoos, Dennis, 193
stream fencing, 39
Strebnicka, Z., 115

strongyle larvae, 76
Sturges, Mark, 191-193, 202-203
sub-atomic particles, 72, 124
succession of plants, 139
Sun God Ra, 179
super-phosphate, 171
"Survival of Horn Flies According
    to the Time of Dung Deposit by
    Cattle, The," 157
swine feces, 72, 124, 125
swine manure, 145-146

Tantalus, 17
temperature, 124
Texas, 159
Texas Bug Book, 157, 159
Tifton Laboratory, 66-67
Tifton, Georgia, 17, 18, 173
Tomorrow's Weather, 33
Tournefort, 117
toxic chemical effects, 173
toxic chemistry, 193
Trox, 190
Trox genus, 147
tumblebugs, 11, 82, 159
Tuning in to Nature, xviii, 10
tunnel system demonstration, 38
tunneler, brood balls, 111
tunneler, chambers, 91
tunneler, copulation, 91
tunneler, facts, 90-91
tunnelers, 82, 138-139
Tutankhamen, 180
Tyndale-Biscoe, Maria, 35

U.S. Entomology Laboratory, 121
urea, 170, 171
USDA and horn flies, 67

USDA curtails research, 66-67
USDA-ARS Biocontrol of Horn Fly
    Unit, 61
USDA-ARS Biological Control of
    Horn Fly Project, 61, 62

Veterinary Parasitology, 155
vindex, 124
Voisin, André, 12

Wallace, Alfred Russell, 56
wasps, 106, 164
water conservation, 39
Watson, James, 57
"We Must Be Rich," 202-203
weakest link, 75
weed control, 138
weeds, 116, 139
whole nine yards, 182
wild onion, 116
Wild Pigs in the United States, 147
Wilkins, Maurice, 57
Wilson, E.O., 50
Wilson, Edward O., 51
wingless scarab, 15
Winter, Will, 188-190
witch beetle, xxi
Woodruff, R.E., 18
wormers, 189

XYp, 107
Y-chromosome, 105, 106, 108

Yucatan Peninsula, 128

Zinjanthropus boisei, 13, 120